Engaging STEM Students From Rural Areas:

Emerging Research and Opportunities

Reenay R.H. Rogers
University of West Alabama, USA

Yan Sun
Mississippi State University, USA

A volume in the Advances in Early Childhood and K-12 Education (AECKE) Book Series

IGI Global
DISSEMINATOR OF KNOWLEDGE

Published in the United States of America by
IGI Global
Information Science Reference (an imprint of IGI Global)
701 E. Chocolate Avenue
Hershey PA, USA 17033
Tel: 717-533-8845
Fax: 717-533-8661
E-mail: cust@igi-global.com
Web site: http://www.igi-global.com

Library of Congress Cataloging-in-Publication Data

Names: Rogers, Reenay R. H., 1965- editor. | Sun, Yan, 1972 June 10- editor.
Title: Engaging STEM students from rural areas : emerging research and opportunities / by Reenay R.H. Rogers and Yan Sun.
Description: Hershey, PA : Information Science Reference, 2019. | Includes bibliographical references.
Identifiers: LCCN 2018008113| ISBN 9781522563419 (hardcover) | ISBN 9781522563426 (ebook)
Subjects: LCSH: Education, Rural--United States. | Rural youth--Education--United States. | Science--Study and teaching--United States. | Mathematics--Study and teaching--United States. | Technology--Study and teaching--United States.
Classification: LCC LC5146.5 .R64 2019 | DDC 372.35--dc23 LC record available at https://lccn.loc.gov/2018008113

This book is published in the IGI Global book series Advances in Early Childhood and K-12 Education (AECKE) (ISSN: 2329-5929; eISSN: 2329-5937)

British Cataloguing in Publication Data
A Cataloguing in Publication record for this book is available from the British Library.

Advances in Early Childhood and K-12 Education (AECKE) Book Series

ISSN:2329-5929
EISSN:2329-5937

Editor-in-Chief: Jared Keengwe, University of North Dakota, USA

MISSION

Early childhood and K-12 education is always evolving as new methods and tools are developed through which to shape the minds of today's youth. Globally, educational approaches vary allowing for new discussions on the best methods to not only educate, but also measure and analyze the learning process as well as an individual's intellectual development. New research in these fields is necessary to improve the current state of education and ensure that future generations are presented with quality learning opportunities.

The **Advances in Early Childhood and K-12 Education (AECKE)** series aims to present the latest research on trends, pedagogies, tools, and methodologies regarding all facets of early childhood and K-12 education.

COVERAGE

- Reading and Writing
- Bullying in the Classroom
- Pedagogy
- Urban K-12 Education
- Special Education
- Learning Outcomes
- Individualized Education
- Head Start and Pre-K Programs
- STEM Education
- Performance Assessment

IGI Global is currently accepting manuscripts for publication within this series. To submit a proposal for a volume in this series, please contact our Acquisition Editors at Acquisitions@igi-global.com or visit: http://www.igi-global.com/publish/.

The Advances in Early Childhood and K-12 Education (AECKE) Book Series (ISSN 2329-5929) is published by IGI Global, 701 E. Chocolate Avenue, Hershey, PA 17033-1240, USA, www.igi-global.com. This series is composed of titles available for purchase individually; each title is edited to be contextually exclusive from any other title within the series. For pricing and ordering information please visit http://www.igi-global.com/book-series/advances-early-childhood-education/76699. Postmaster: Send all address changes to above address.

Titles in this Series

For a list of additional titles in this series, please visit:
https://www.igi-global.com/book-series/advances-early-childhood-education/76699

Early Childhood Education From an Intercultural and Bilingual Prspective
Cristina A. Huertas-Abril (University of Córdoba, Spain) and María Elena Gómez-Parra (University of Córdoba, Spain)
Information Science Reference • ©2018 • 302pp • H/C (ISBN: 9781522551676) • US $190.00

Handbook of Research on Positive Scholarship for Global K-20 Education
Victor X. Wang (Grand Canyon University, USA)
Information Science Reference • ©2018 • 388pp • H/C (ISBN: 9781522556671) • US $245.00

Optimizing Elementary Education for English Language Learners
Nilufer Guler (Avila University, USA)
Information Science Reference • ©2018 • 335pp • H/C (ISBN: 9781522531234) • US $195.00

Curriculum Integration in Contemporary Teaching Practice Emerging Research and ...
Susannah Brown (Florida Atlantic University, USA) and Rina Bousalis (Florida Atlantic University, USA)
Information Science Reference • ©2018 • 193pp • H/C (ISBN: 9781522540656) • US $130.00

Data Leadership for K-12 Schools in a Time of Accountability
Evan G. Mense (Southeastern Louisiana University, USA) and Mindy Crain-Dorough (Southeastern Lousiana University, USA)
Information Science Reference • ©2018 • 332pp • H/C (ISBN: 9781522531883) • US $195.00

Developing Effective Literacy Intervention Strategies Emerging Research and Opportunities
Cheryl A. Slattery (Shippensburg University, USA)
Information Science Reference • ©2018 • 145pp • H/C (ISBN: 9781522550075) • US $125.00

Instructional Strategies in General Education and Putting the Individuals With ...
Pam L. Epler (Grand Canyon University, USA)
Information Science Reference • ©2018 • 344pp • H/C (ISBN: 9781522531111) • US $175.00

For an entire list of titles in this series, please visit:
https://www.igi-global.com/book-series/advances-early-childhood-education/76699

IGI Global
DISSEMINATOR OF KNOWLEDGE

701 East Chocolate Avenue, Hershey, PA 17033, USA
Tel: 717-533-8845 x100 • Fax: 717-533-8661
E-Mail: cust@igi-global.com • www.igi-global.com

Table of Contents

Foreword vii

Preface ix

Acknowledgment xv

Chapter 1
STEM Education in Rural Areas: From Research to Reality 1

Chapter 2
Project Engage: Solutions for Recruiting, Retaining, and Reinforcing STEM Majors From Rural Areas 21

Chapter 3
Scaffolding Undergraduate STEM Majors: A Strategic Mentoring Program 39

Chapter 4
A Three-Pillar Approach to Preparing Tomorrow's STEM Professionals: Developing Knowledge, Abilities, and Ways of Working 64

Chapter 5
Opening the Doors to STEM Employment Opportunities for Rural Youth: Guided Career Exploration 120

Chapter 6
Improving Undergraduate STEM Education: A Four-Dimensional Framework 163

Related Readings 182

About the Authors 203

Index 204

Foreword

Enrollment and persistence of U.S. students in science, technology, engineering, and mathematics (STEM) disciplines at the post-secondary level fail to meet the nation's current need to fill vacancies in the STEM workforce. The "pipeline" from post-secondary education to STEM career suffers from both low intake (i.e., low post-secondary enrollment) and "leaking" (i.e., changing majors, dropouts, and failure to enter the STEM workforce). Drs. Rogers and Sun, in collaboration with colleagues at The University of West Alabama and education and business stakeholders in the Black Belt region of the state, and with major federal grant funding, have worked extensively in rural STEM education initiatives. The authors have gained expertise in identifying, understanding, and putting solutions into place to increase the efficiency of the rural STEM pipeline from high school/college to career. This book examines solutions to increase the intake of U. S. students into the STEM pipeline as well as methods to stop the leaks at the post-secondary level, thereby creating a steady flow of students through the pipeline.

One solution to increasing the pool of candidates going into the pipeline is to capitalize on an often-overlooked classification of students (i.e., rural students). Rural school districts constitute a large percentage (57%) of American school districts, yet efforts to recruit from this large pool of students into STEM disciplines are absent from the literature. The literature is replete in regards to STEM education in general; but deficient of information on rural STEM education. The time has come to take a closer look at rural schools and rural school students as a means to supply and diversify the U. S. STEM workforce. The authors review the common barriers to post-secondary education such as poverty, lack of funding, lack of educational resources and inadequate STEM preparation, low parental expectations, etc., often faced by rural students. Overcoming these barriers is the first step to supplying the STEM pipeline.

In general, students in post-secondary STEM education far too often "leak" from the STEM pipeline. This is especially true for rural students who are frequently the victims of unequal access to rigorous STEM education, STEM professional role models, and opportunities to explore STEM careers due to their isolated geographic locations. Drawing upon their research and professional experiences, Drs. Rogers and Sun propose a comprehensive, research-based approach for addressing the leakage, developed as part of a Minority Science and Engineering Improvement Program grant funded by the U.S. Department of Education (Project Engage). The primary components of Project Engage include three tiered strategic mentoring developed from a social constructivist perspective, STEM engagement activities for motivation and developing the knowledge and skills needed as a STEM professional, and guided STEM career exploration in order to improve STEM career self-efficacy. Together these three major components are incorporated into an overarching model, the Improving STEM Education Framework centered around STEM experiences, engagement, and scaffolding to stay motivated to complete an undergraduate program.

Supplying the nation's STEM workforce needs depends on identifying new pools of candidates to fill the STEM pipeline as well as methods to prevent these candidates from leaking out of the pipeline before entering the workforce. The authors have identified a previously untapped group of the nation's students, rural school students, as having great potential. Through concentrated efforts such as those of Project Engage, barriers to post-secondary education can be addressed and STEM career aspirations developed.

Susan Cordell
University of West Alabama, USA

Preface

INTRODUCTION

The term “rural America” is deceptively simple; it is simple in the sense that “rural” seems perpetually associated with the mental images of sparsely populated areas with sprawling expanses of farmland and pasture, but it is deceptive because, with this single term, we are talking about 97% of the country’s land mass and home to about 60 million people (United States Census Bureau, 2017). When it comes to rural education, this word “rural” applies to more than 49,500 public elementary and secondary schools serving about 12 million students. Despite the diversity of rural areas in terms of their geographic features, natural environment conditions, economic structures, and population makeup, public schools in rural areas are facing similar obstacles or impediments such as limited educational resources, difficulties in recruiting quality teachers, poverty and lack of funding, low educational attainment and aspirations in students, and low parental awareness and expectations. These unfortunate similarities among rural schools defy such comments as “there are no two identical rural areas” and “each rural area is unique”.

Improving STEM (science, technology, engineering, and mathematics) education and strengthening the STEM workforce pipeline have long been acknowledged as national priorities. Incessant efforts have been made to address these national priorities through educational research, innovative STEM education initiatives, and teacher professional development. However, rural schools and students have been largely left out of such efforts. The aforementioned similarities among rural schools make them a less desired spot on the STEM landscape. With 12 million students in rural schools, rural areas represent one of the greatest, yet underexploited, opportunities to impact STEM workforce development. Educational programs are needed to realize such impact. Project Engage, introduced in this book, is one of such educational programs that aimed to strengthen the STEM pipeline running from college to STEM career.

PROJECT ENGAGE

While Chapter 1 outlines and expounds on the barriers in rural education, the educational gaps in rural students, and the potential of rural schools to impact STEM workforce pipeline, Chapter 2 gives a detailed introduction of Project Engage as an educational program for preparing rural undergraduate students from the Alabama Black Belt region, particularly rural minority students, for STEM futures. Project Engage is a grant initiative funded by the Minority Science and Engineering Improvement Program (MSEIP) of the U.S. Department of Education. In addition to an overview about the MSEIP grant and Project Engage's overall goals and objectives, Chapter 2 elaborates on the STEM education barriers Project Engage is intended to address including poor educational preparation, lack of STEM experience, inadequate technology preparation, lack of motivation, and lack of professional role models. Project Engage activities, designed and implemented to address these barriers, fall into three categories: strategic mentoring, STEM engagement, and guided career exploration. A summary of activities in each of the three categories is presented in Chapter 2 but detailed information of these activities is revealed in Chapter 3, Chapter 4, and Chapter 5.

SCAFFOLDING UNDERGRADUATE STEM MAJORS

Chapter 3 introduces the Scaffolding model developed and adopted by Project Engage to guide its strategic mentoring activities for STEM undergraduate students. The Project Engage Scaffolding model embraces a constructivist view of learning and the scaffolding theory based on More Knowledgeable Other (MKO) and Zone of Proximal Development (ZPD). Guided by this Scaffolding model, Project Engage implemented a three-tiered strategic mentoring program consisting of peer mentors, a career counselor, and STEM faculty. This hierarchical community of mentors serve as MKOs guiding students through their first year as college freshmen majoring in a STEM field. In addition, Chapter 3 also provides detailed information about how the three mentoring tiers function, how peer mentors are selected and trained, and what are the respective responsibilities of peer mentors, the career counselor, and faculty mentors.

The "Research Findings From a Strategic Mentoring Program" section in Chapter 3 presents the research with survey data collected from 2012, 2013, and 2016 Project Engage student cohorts. The surveys were intended to reveal the overall effect of the strategic mentoring program and to measure its impact on undergraduate STEM students' academic performance, persistence as STEM majors, and intention to choose a STEM career after graduation. The results of the survey data analysis are reported and future research directions based on the data analysis results are also provided in Chapter 3.

PREPARING TOMORROW'S STEM PROFESSIONALS

STEM professionals need specific knowledge, abilities, and general ways of working to be successful. Chapter 4 presents the three-pillar approach for preparing tomorrow's STEM professionals adopted by Project Engage in developing the STEM engagement activities for undergraduate STEM students. The three pillars are:

- Knowledge of STEM careers and professionals;
- Abilities to work in the STEM fields; and
- Ways of working as STEM professionals.

Chapter 4 also gives detailed descriptions of the activities comprising the three pillars of preparation. These activities include the following:

- Professional exploration trips
    - U. S. Space and Rocket Center
    - Alabama Nature Center
    - Marshall Space Flight Center/Redstone Arsenal
    - HudsonAlpha Institute for Biotechnology
    - Mercedes-Benz U.S. International Inc.
- Competitive STEM Events
    - Invention Convention: Robotics Year 1
    - Invention Convention: Robotics Competition with the Experts
    - STEM Challenge: Forensic Year 1
- Invited STEM Speakers
- Team Building Training

In the "Research Findings From Activities of Preparing Tomorrow's STEM Professionals" section, Chapter 4 reports the research with data from field trip surveys and from STEM challenge and invention convention surveys. The survey data analysis results reveal the impact of the activities on undergraduate STEM students' competency to be successful in STEM, knowledge of STEM fields, skills of working on STEM projects, persistence as STEM majors, and intention to work as STEM professionals after graduation. In addition, the results from analyzing the team building activity survey data are reported and the impact of team building activities on students' team cooperation and personal confidence are revealed. Based on the survey data analysis results, Chapter 4 also furnishes STEM educators and STEM educational researchers with future research directions.

GUIDED CAREER EXPLORATION

Rural high schools often lack the resources for providing career counseling services and rural students often lack realistic visions of what STEM careers entail as their experience with STEM professionals is limited. Being aware of such gaps, Project Engage adopts a tripartite approach for guided career exploration. Chapter 5 provides an overview of this tripartite approach of career decidedness, career thoughts, and vocational maturity and specifies the aims to be achieved by this approach as 1) enhance STEM students' career decidedness by providing STEM career information, having students study their own interests, and exploring STEM career opportunities; 2) cultivate positive STEM career thoughts by increasing students' self-efficacy about STEM careers and by identifying, challenging, and altering students' negative STEM career thoughts; and 3) enhance students' vocational maturity by involving students in STEM career planning and development activities to improve their awareness of what is required to make an appropriate career choice and their ability to make career decisions that are both realistic and consistent over time.

Chapter 5 also provides a description of guided career exploration activities for Project Engage students that include:

- Career exploration course;
- Career panel discussion;
- Undergraduate research; and
- Sisters of the Academy "Priming the Pipeline."

Project Engage realizes that a supply of students is needed to increase enrollment in STEM majors and to ultimately increase the flow of STEM professionals into the workforce. Outreach to high schools is needed to ignite student interest in STEM during the secondary education years in order to increase enrollment in postsecondary STEM programs. Therefore, Project Engage includes an outreach component to introduce high school students to STEM majors and careers. Chapter 5 provides information about the following high school student outreach activities in Project Engage:

- Green Convention;
- Shark Tank Competition;
- Boomerang Convention;
- Building a Better Invention;
- STEM Challenge: Forensics;
- STEM Challenge: S-T-E-M Round Robin; and
- STEM Challenge: Return to Forensics.

A career panel discussion survey was conducted among the 2013 Project Engage Cohort. The survey data and data analysis results are presented in Chapter 5. The results reveal the effects of the Career Panel Discussion on the undergraduate students' interest in STEM fields, knowledge of STEM fields, persistence as STEM majors, and intention to work as STEM professionals after graduation. In addition, Chapter 5 also presents the data and data analysis results from high school 2013, 2014, and 2015 invention convention surveys and from high school 2013, 2014, and 2016 STEM challenge surveys. The data analysis results reveal the impact of the Project Engage outreach activities on high school students' interest in STEM fields, knowledge of STEM fields, intentions to enroll as STEM majors, and intentions to work as STEM professionals after college. Future research directions are pointed out in Chapter 5 based on the survey research results reported in the chapter.

IMPROVING UNDERGRADUATE STEM EDUCATION

Educational environments in rural areas are characterized by high levels of poverty, less available cultural capital and enrichment activities outside of the regular school curriculum, lack of qualified STEM teachers, lack of role models of successful STEM professionals, and inadequate awareness of STEM career opportunities. As typical rural STEM majors growing up in such an educational environment, Project Engage students are not well prepared for academic success in the learning of their STEM disciplines and often lack the motivation to complete their STEM degree and pursue a STEM career after graduation. Project Engage designed and implemented programs and activities to achieve the ultimate goals of improving rural undergraduate STEM programs' capacity of retaining undergraduate students in STEM majors, cultivating STEM talents, and producing motivated and dedicated STEM graduates. Survey research reported in Chapters 3-5 has indicated positive results of Project Engage programs and activities in achieving the project goals. Based on its successful experience in these programs and activities, Project Engage developed a four-dimensional practice-and research-based framework that can be used to guide future efforts of improving undergraduate STEM education in rural areas. This framework is presented in Chapter 6.

The four dimensions in the Improving Undergraduate STEM Education Framework are: 1) motivating, 2) experiencing, 3) engaging, and 4) scaffolding. Motivation to pursue STEM degrees and careers is the key to improving undergraduate STEM education in rural areas. The *Motivating* dimension in the framework centers on improving undergraduate STEM students' intrinsic motivation through the following aspects: improving undergraduate students' self-efficacy in STEM learning; increasing their interest in STEM disciplines and STEM careers; and enhancing their ability to work as STEM professionals. The other three dimensions (i.e., *Experiencing*, *Engaging*, and *Scaffolding*) in the Improving Undergraduate STEM Education Framework serve as means to achieve the ends of the *Motivating* dimension.

In addition to a detailed description of the four dimensions in the Improving Undergraduate STEM Education Framework, Chapter 6 includes a revisit of the survey data analysis results reported in Chapters 3-5 providing evidence that this framework is solidly grounded in the practice and research of Project Engage and can serve as a valid and useful tool for guiding STEM educators and STEM educational researchers in their design and development of future interventions or initiatives aimed to improve undergraduate STEM education in rural areas. Limitation of the framework and implications for future research are also discussed in this chapter.

REFERENCES

United States Census Bureau. (2017). *What is Rural America?* Retrieved from https://www.census.gov/library/stories/2017/08/rural-america.html

Acknowledgment

We would like to acknowledge the contributions of the Project Engage faculty, staff, mentors and student participants. Our fellow colleagues, Dr. Ketia Shumaker, Dr. Frank Rogers, and Dr. Balu Gokaraju, have worked alongside us as part of the Project Engage Leadership Team. Their knowledge and skills have supported Project Engage's effort to impact the lives and STEM careers of students form the Black Belt of Alabama. It is our hope that the Project Engage career counselors, mentors, and mentees will become role models for future generations of students from this rural area.

We would also like to acknowledge our families for their never-ending support of our efforts to promote STEM education. It is our belief that through improved STEM education, we can create a citizenry capable of making informed decisions in this ever-advancing technological world.

Chapter 1
STEM Education in Rural Areas:
From Research to Reality

ABSTRACT

Science, technology, engineering, and mathematics (STEM) education is challenged to produce enough graduates to maintain nationwide STEM workforce needs. This chapter discloses the importance of STEM education for developing a more scientifically literate citizenry prepared to face the challenges of an increasingly technological society and meet U.S. workforce demands. An examination of the state of STEM education reveals its current inability to produce enough qualified graduates to meet current and future needs. The researchers propose looking towards rural areas to supply the STEM pipeline running from college to STEM career. In order to capitalize on the rural school population, programs are needed to address the barriers between the rural student and college. This chapter outlines and expounds these barriers as well as introduces Project Engage as one such program determined to prepare rural students from the Alabama Black Belt region, particularly rural minority students, for STEM futures.

INTRODUCTION

STEM education has been at the forefront of research as investigators seek new ways to address shortages of STEM professionals and technical workers. Low performance of U. S. students in science and mathematics has only exacerbated the issue, highlighting the need for improved STEM education.

DOI: 10.4018/978-1-5225-6341-9.ch001

This chapter introduces rural students as a source for supplying the STEM pipeline. Rural students may be the solution to increasing the number of students entering postsecondary education with a STEM major and ultimately adding to the STEM workforce; however, rural students face many barriers to college entrance and completion such as limited educational resources, STEM teacher shortages, poverty, lack of school funding, and low parental awareness and expectations. Thus, a secondary purpose of this chapter is to identify and describe the impediments rural students face. Finally, the researchers will introduce Project Engage, a Minority Science and Engineering Improvement Program (MSEIP) grant initiative, designed to support the needs of rural minority students, specifically those from the rural Black Belt area of Alabama in pursuit of a bachelor's degree in a STEM field.

STEM IMPORTANCE TO SOCIETY

There has long been a broad consensus that continued U. S. competitiveness in an increasingly global economic environment relies heavily on the adequacy of supply and the quality of the workforce in the STEM fields (National Research Council, 2011; President's Council of Advisors on Science and Technology, 2010; U.S. Department of Labor, 2007). The National Academies *Gathering Storm* committee (National Academy of Sciences, National Academy of Engineering, and Institute of Medicine, 2011) concluded:

A primary driver of the future economy and concomitant creation of jobs will be innovation, largely derived from advances in science and engineering. While only 4 percent of the nation's workforce is composed of scientists and engineers, this group disproportionately creates jobs for the other 96 percent (National Science Board, 2014).

THE STATE OF STEM EDUCATION

Knowledge of STEM is not just required of professional scientists and engineers. It is now required for an increasing number of jobs at all levels. It is a national advantage to have a more diverse STEM pipeline and more STEM literate citizens. However, disconcerting statistics give us good reasons to be concerned about STEM education. The National Assessment of Educational Progress (Schmidt, 2011) reported about 75% of U. S. 8th graders are not proficient in mathematics. U. S. students lag behind highest performing nations on international assessments: only 10% and 12% of 8th graders respectively reached the advanced international benchmark in mathematics and in science in 2015 and the percentages for 4th graders reaching mathematics and science benchmarks were 14% and 16% (National Center for Educational Statistics, 2015).

On top of the overall underperformance in STEM disciplines, there are significant achievement gaps among student population groups. For example, according to statistics from the National Center for Educational Statistics and the NSF Division Resources Statistics (National Research Council, 2011), white, Hispanic, and black students respectively, have a gain of 116 points, 113 points, and 101 points in average mathematics scores from kindergarten to 8th grade. By 5th grade, the gap between white and black students in average mathematics scores was 24 points, and the average score of black 5th grade students was equivalent to the average 3rd grade score of white students. Similar mathematics achievement gaps persist through high school (Ingels, Pratt, Rogers, Siegel, & Stutts, 2004; Riegle-Crumb & Grodsky, 2010).

National math and science performance scores are indicative of a STEM education system that is failing to prepare a strong U. S. workforce with the knowledge and skills needed for success (Allendoerfer, Wilson, Kim, & Burpee, 2014; National Academy of Sciences, 2012). Currently, K-12 education "emphasizes discrete facts with a focus on breadth over depth, and does not provide students with engaging opportunities to experience how science is actually done" (National Academy of Sciences, 2012, p. 1). The National Academy of Sciences also acknowledges a need for new approaches that will create informed citizenry capable of choosing and succeeding in careers in STEM. Likewise, STEM education at the undergraduate level has been, and for the most part remains, teacher-directed with a knowledge and assessment-centered environment and little or no attention given to the learner or community-centered environments. STEM faculty continue to focus on content delivery rather than student inquiry (Allendoerfer, Wilson, Kim, & Burpee, 2014).

The Next Generation Science Standards (NGSS) for K-12 students were developed in an attempt to move away from teacher directed instruction and rote memorization towards fostering a more in-depth understanding of scientific principles and key STEM skills such as collaboration, communication, and problem-solving. The outcome of implementation of NGSS is yet to be seen; however, expectations are high that the new standards will cultivate a more scientifically literate citizenry with a desire to pursue post-secondary studies in a STEM field (NGSS Lead States, 2018). While there are no Next Generation Science Standards for post-secondary STEM education, the need for programs which inspire and support students as they progress through STEM degree programs at this level is a necessity that can no longer be overlooked.

THE STEM PIPELINE

The STEM pipeline, as a theoretical construct, describes the flow of students through the K-12 education system and the higher education system toward employment in STEM fields. Gender, race, and socio-economic status (SES) are all influences on student choice to major in a STEM field and success. Higher SES tends to positively affect the choice by minorities and females to major in STEM fields. However, students from lower SES backgrounds often make decisions to major in STEM fields ill-suited to their math skill level leading to lower success rates (Niu, 2017). Furthermore, U. S. students are discouraged from entering the STEM pipeline due to either inadequate preparation in math and science or poor teacher quality in their K-12 education (American College Testing Program, Inc., 2005, 2006). Unsurprisingly, international students constituted more than a third of the students in U.S. science and engineering graduate schools (National Academy of Sciences, National Academy of Engineering, and Institute of Medicine, 2007). A study at a Midwestern university found that 42% of students who enrolled in the College of Science and Mathematics left college after the freshman year (Prescod, 2014), and this is not uncommon in other U. S. universities. In the field of engineering, the trends in undergraduate and graduate enrollment in engineering science are troubling (U.S. Government Accountability office, 2005). If such trends continue, as warned by the Business Roundtable (2005), more than 90% of all scientists and engineers in the world will live in Asia.

THE RURAL SOLUTION

The phrase "rural schools" brings to mind pictures of small, geographically isolated schools, yet this definition lacks specificity needed for research. To clarify "rural," the National Center for Education Statistics revised its definition of U. S. locales in 2006 based on a collaboration with the U.S. Census Bureau to reflect specific geographic locations. Knowing the classification of these locales allows for a more precise definition of rural schools (i.e. schools within the rural locale). The geographic classification identifies four primary locales (i.e. city, suburb, town, rural) which are then sub-divided into three more specific areas. Rural areas are classified as either fringe, distant, or remote based on their distance from an urbanized area or urbanized cluster: 1) less than or equal to 5 miles from an urbanized area, as well as rural territory that is less than or equal to 2.5 miles from an urban cluster (fringe); 2) more than 5 miles but less than or equal to 25 miles from an urbanized area, as well as rural territory that is more than 2.5 miles but less than or equal to 10 miles from an urban cluster (distant); and 3) is more than 25 miles from an urbanized area and is also more than 10 miles from an urban cluster (remote) (National Center for Education Statistics, 2006). Although rural schools have been defined by geographic classifications, there still exists "diverse cultural, social, economic, and political realities that make rural places what they are today" (Williams, 2010).

It is a national priority to improve STEM education and to strengthen the STEM workforce pipeline. Although the literature is replete with educational research intended to help address this national priority, a strictly geographical perspective focusing on rural students has been largely invisible on the STEM education research landscape. Rural students constitute a significant portion of the U. S. public school population. The National Center for Education Statistics (2013) reported "In 2010–11, more than half of all operating regular school districts were located in rural areas (57 percent), while 20 percent of districts were located in suburban areas, 18 percent in towns, and 5 percent in cities" (p.1). The high percentage of students from rural school districts points to an ample population for targeted recruitment and retention efforts to fill the gaps in the STEM workforce.

Rural school districts prepared over 12 million public elementary and secondary students in 2015-2016. More recent statistics provided by Showalter, Klein, Johnson, Hartman, and The Rural School and Community Trust (2017) reveal:

- Almost 8.9 million children attend schools classified as rural;
- More than 25% of America's schools are rural;
- Over one in six of school-aged children attend rural schools;
- More than one in four rural students is a child of color; and
- Almost 50% of the students in rural schools are from low-income families.

However, despite these astounding numbers, rural students have been an underrepresented population in STEM education research and there are various obstacles in the rural pathways into STEM fields preventing rural students from playing an important role in diversifying and strengthening STEM workforce pipeline.

BARRIERS FACED BY RURAL STUDENTS

An examination of current literature identifies several impediments for students from rural areas attending post-secondary institutions (Byun, Meece, & Irvin, 2012) and eventually entering the STEM workforce. Such obstacles include fewer advanced course offerings in mathematics and science (Babco, 2003; Carsey Institute, 2009), education by STEM teachers who are less prepared than their urban counterparts (Fowler et al., 2013; Gagnon & Mattingly, 2012), as well as less funding for K-12 schools (Versypt & Versypt, 2013). More intangible but equally impactful factors involve high poverty levels, lack of parental participation and expectations (Provasnik et al., 2007), and lack of exposure to STEM professionals serving as role models (Pitchford, 2016; Versypt & Versypt, 2013). These combined factors negatively impact rural students' pathway to STEM careers.

Gap in College/University Enrollment

The 2015 report from the National Center for Educational Statistics shows that the percentage of 18-24 year olds enrolled in colleges or universities was 29.3% at rural locales (a decline from 31.3% in 2008), 42.3% at city locales, 47.7% at suburban locales, and 41.2% at town locales. This lower college/university enrollment rate at rural locales is confirmed in a 12-year longitudinal study on rural-nonrural disparities in postsecondary educational attainment indicating that rural students were less likely to enroll in colleges or universities (Byun et al., 2012). Although statistics of rural STEM students are unknown in the literature, the low and declining college or university enrollment rate of rural youths give rise to concerns about rural students' participation in STEM fields.

Limited Educational Resources

Rural students have fewer educational resources as compared to students in other locales (Avery, 2013; Boynton & Hossain, 2010), and this might be a factor hindering them from entering the STEM pipeline. Typically, there are fewer advanced courses offered in rural areas (American Enterprise Institute, 2016; Anderson & Chang, 2011; Guckian & Sarrio, 2012; Carsey Institute, 2009). Limited access to advanced mathematics and science courses adversely impacts the number of qualified rural students entering STEM job pipelines (Babco, 2003; Carsey Institute, 2009). Inadequate educational resources for rural students are also evident by the limited access to educational resources offered by colleges, business, and other educational organizations (Avery, 2013); lack of access to STEM professionals as mentors or role models (Pitchford, 2016; Versypt & Versypt, 2013); and few STEM outreach opportunities (Boynton & Hossain, 2010; Versypt & Versypt, 2013). Irrespective of higher costs for obtaining services in rural areas due to a large portion of smaller schools, "only 17% of state education funding goes to rural districts" (Showalter et al., 2017, p. 1). Within the rural geographic and cultural context, students lack access to STEM professionals as mentors or role models (Pitchford, 2016; Versypt & Versypt, 2013). This could prevent student awareness of STEM career opportunities resulting in "leaking" from the STEM pipeline as students may not have interest in pursuing degrees in fields to which they have little or no exposure (Versypt & Versypt, 2013).

Difficulties Recruiting STEM Teachers

Schools in rural areas find it difficult to recruit and retain teachers, particularly in science, mathematics, and special education (Pitchford, 2016; Schafft, 2016; Versypt & Versypt, 2013). This may explain why rural schools tend to have a larger number of inexperienced teachers than urban or suburban schools (Fowler et al., 2013; Gagnon & Mattingly, 2012). In general, STEM teachers in rural schools are responsible for teaching many different subjects and find themselves a long distance from any professional support networks (Marder, Brown, & Polisch, 2017). In addition, STEM teachers in rural schools are less likely than their peers in other locales to have majored in science or have a graduate degree. Adding to the already complex issue, STEM teachers also have fewer opportunities for professional development (Hossain & Robinson, 2012). Rural students, therefore, have fewer options of STEM courses offered by qualified STEM teachers. Engberg and Wolniak (2013) concluded "Course-taking, in particular, proved to be an extremely important factor in raising STEM propensities" (p. 27). Lacking the teacher manpower and subsequently a lack of course offerings in advanced STEM subjects negatively impact rural students in their pursuit of STEM majors in college.

Poverty and Lack of Funding

Rural areas are likely to be the most impoverished. According to Versypt and Versypt (2013), 244 of the 250 poorest counties in the U.S. were rural. A large percentage of rural students live in poverty (e. g. 39% in Illinois and 30% in Kansas) (Beeson & Strange, 2000). Compounding the issue of poverty is the issue of dispersion: small, scattered, and isolated school districts. Dispersion, together with an impoverished population, leads to underfunding (Strange, 2011) as this combination makes rural schools less likely to capture the attention of legislators dispensing the state educational funds. Poverty and lack of funding make teacher salaries in rural schools lower than national and state averages. This, at least partially, explains why rural schools have trouble of recruiting and retaining teachers, particularly in science, mathematics, and special education (Pitchford, 2016; Schafft, 2016; Versypt & Versypt, 2013). Affluent suburban school districts hire the most qualified teachers, while poorer rural school districts have to hire a larger number of underqualified STEM teachers or none at all (Aragon, 2016; Fowler et al., 2013; Gagnon & Mattingly, 2012; Mader, 2014; Ossola, 2014). Lack of funding also widens the digital divide between rural schools and their urban counterparts. Many rural schools lack the needed computer hardware and sufficient bandwidth for online education delivery (Smith, 2017). Underfunding perpetuates inequities encountered and low teacher salaries (Williams, 2010). Poverty and lack of funding undermine rural students' academic performance, limit their awareness of and exposure to STEM career opportunities, and lower their educational aspirations.

Low Parental Awareness and Expectation

Fewer parents in rural areas have college degrees as compared to city, suburban, and town parents. The report, Status of Education in Rural America by the Center of Educational Statistics (Provasnik et al., 2007) shows that, in 2004, a larger percentage of school-age children in rural areas had a mother or father who completed high school as their highest level of educational attainment than their peers in cities and suburbs. Keeping with the trend of lower levels of educational attainment, the percentage of school-age children with a mother or father with a bachelor's degree as their highest attainment was lower in rural areas than in suburban areas. Lower parental educational attainment has made rural parents less aware of STEM careers resulting in lower educational expectations of their children. According to the same report by the Center of Educational Statistics (Provasnik et al., 2007), the percentage of rural students whose parents expected their highest educational attainment to be less than a bachelor's degree (42 percent) was larger than the percentages of students in cities and suburban areas (30 and 25 percent, respectively). Furthermore, rural students aspire to what they know; thus, when their parents are in agricultural or manual jobs, those are the occupations that they know and those are the careers they tend to pursue (Haller & Virkler, 1993).

RURAL AREAS: THE GREATEST OPPORTUNITY FOR STEM EDUCATION TO IMPACT WORKFORCE DEVELOPMENT

With the previously mentioned problems and obstacles that affect rural public K-12 systems, it comes as no surprise that rural students have low educational aspirations and are not well prepared for entering the STEM pipeline. It has been realized that efforts must be made to increase the availability of rigorous courses, including AP courses, in remote rural areas because it is these more rigorous math and science courses that provide the building blocks for STEM careers (Babco, 2003). Research has shown that undergraduate STEM students from rural areas are less academically prepared and have lower GPAs than their urban and suburban counterparts. Not only being less prepared for STEM, rural youths have lower level of educational aspirations than non-rural youths. Rural students aspire to different types of careers than non-rural students: while non-rural students aspire to professional jobs, rural students often lean towards occupations that require less education, typically in agriculture, service industries, and manufacturing (Versypt & Versypt, 2013).

Despite the lower educational aspirations and inadequate educational preparedness found among rural students, rural schools have the greatest potential for STEM education to impact workforce development. The new school locale classification system revised by the National Center for Education Statistics (NCES) in 2006 specifies four major locale categories: city, suburban, town, and rural. In the school year 2010-11, some 99,000 public elementary and secondary schools, located in 14,000 school districts, served over 49 million students in the United States (National Center for Educational Statistics, 2013). More than half of these school districts are located in rural areas serving about 24% of the 49 million students (see Figure 1).

Although rural schools constitute about one third of all public schools (National Center for Educational Statistics, 2013), researchers and lawmakers have largely focused on urban and suburban schooling. It is important to recognize that rural areas represent one of the greatest, yet underexploited, opportunities for STEM education to impact workforce development (Carnegie Science Center, 2014). The importance of rural STEM education to impact workforce development lies not only in the large rural student population but also in the greatest opportunity for STEM related careers in new industries at rural areas (Carnegie Science Center, 2014). Several STEM initiatives have been launched in recent years in rural areas (e.g., Texas STEM Degree Accelerator Program, Rural Communities STEM Initiative, Project Engage). A common characteristic of such initiatives is dealing with rural education challenges to improve STEM education and to strengthen the STEM workforce pipeline in rural areas. Project Engage, targeting minority students from a rural area of the U.S. known as the Black Belt Region, is one of such initiatives.

Figure 1. Percentage distribution of public elementary and secondary students, schools, and districts, by locale: School year 2010-11 (National Center for Educational Statistics, 2013)

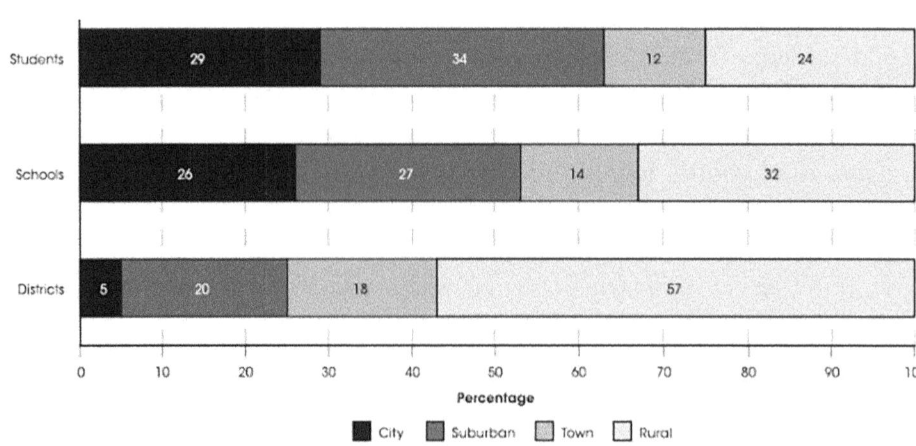

Project Engage is a program funded by a U. S. Department of Education Minority Science and Engineering Improvement Program grant which seeks to increase the number of underrepresented groups in STEM fields. Through funding by this grant, Project Engage is poised to motivate and prepare rural, minority students towards future STEM careers.

The Environment of Counties in the Black Belt Region

The Black Belt includes an area which stretches in a crescent shape from lower Arkansas through central and west Alabama into Georgia. However, Project Engage serves only the portion of the Black Belt Region in Alabama. This region includes the poorest counties in Alabama and the United States. Arthur Raper (1936) noted that the Black Belt contained some of "the richest soil and poorest people" in the United States. The name, Black Belt, refers to the thin layer of exceptionally fertile black soil that encouraged cotton farming in the pioneer period of Alabama history. Today, the major characteristics of Black Belt counties include: low rates of educational attainment, lack of economic opportunity, and the highest levels of poverty in the state and nation.

Alabama ranks third on Showalter et al.'s (2017) ranking of rural schools with the greatest challenges. The ranking considers five gauges when scoring states: importance of rural education in the state, diversity of rural students and their families, educational policy context for rural schools, educational outcomes of rural students, and college readiness of students in rural schools (2017). Factors attributing to the ranking of Alabama include the following.

- There are 265,000 rural students with the majority from low-income families (six out of ten).
- Educational spending at approximately $4,800 per student is one of the lowest amounts in America.
- Teacher salaries are among the lowest in the nation, with the average under $50,000.
- Students in 4th and 8th grade have some of the lowest mathematics scores in America.
- Rural high school graduation rates are below the nation's average.

One factor contributing to Alabama rural schools' need for improvement is economically related. For example, "Of the 15 counties in Alabama with double-digit unemployment, 14 are rural" (Lee & Summers, 2003, p. 5). Furthermore, "of the 20 counties with the lowest median family income, all are rural" (Lee & Summers, 2003, p. 5). The counties of the Black Belt area are rural in nature and under-developed with few employment opportunities due to a lack of a diversified economy, a lack of qualified workforce, and a lack of infrastructure. Such employment hardship evolves from several factors. The rural environment of the Black Belt is taxed by economic extremes: high levels of poverty, low standards of living, high unemployment rates, and low levels of funding for schools. Extreme academic deficiencies also impact the potential of the region's children to ultimately pursue STEM careers: poor achievement scores of K-12 students, high percentages of teachers teaching out of their fields, high dropout rates, and low numbers of high school graduates attending post-secondary institutions (Alabama School Connection, 2013; Alabama State Department of Education, 2013-2014; Kids Count for Alabama Data Book, 2013). These circumstances shape the education and ultimately the futures of the children living in the region.

The nine county school systems in the Project Engage service area are characterized by extreme poverty and low academic achievement. A majority of the students come from disadvantaged backgrounds as documented by the high percentage on free/reduced lunch (51% to 97%), a high number of families living below the poverty level (24.5% to 38.7%), and 100% Title I schools. The ACT scores of students in the nine Black Belt county public school districts have an average less than 15 (see Table 1). Within the nine Black Belt counties in the Project Engage service area, low academic performance is the status quo. Adding to the academic characteristics is an average dropout rate of 13%, graduation rate of 87%, and, most pointedly, a low percentage (11%) of students attending a post-secondary institution. The significance of this data represents an ongoing cycle of the poor educational standards in which students have no exposure to a quality education, thus the few who complete college and return do not change the existing norm established decades ago.

Rural students are a previously untapped population poised to fill the STEM pipeline; however, steps must be taken in order to adequately support their needs as they progress through college. Project Engage, recognizing the unique cultural traditions of the region, along with the natural resources that make the Black Belt a region rich with potential research, economic, and social opportunities, has been making efforts to deal with STEM educational challenges for the purpose of increasing the quantity and quality of STEM talents in the region. Through STEM engagement activities, STEM skills

Table 1 K-12 School demographic information for West Alabama counties

	Choctaw	Clarke	Greene	Hale	Marengo	Pickens	Perry	Sumter	Wilcox
Enrollment	1639	3050	1186	2749	1253	2699	1637	1755	1800
% Free/ Reduced Lunch	81%	50.8%	94.5%	77.9%	78.7%	74.1%	72.8%	96.5%	73.29%
% Title 1 Schools	100%	100%	100%	100%	100%	100%	100%	100%	100%
% African American Students	66.4%	49.9%	76.6%	69.3%	68.4%	59.7%	64.4%	98.4%	93.7%
% Drop-Out	18%	13%	15%	20%	7%	12%	7%	10%	13%
Per Student Expenditure	$8,968 .	$8,587	$10,702	$8,602	$9,292	$9,035	$9,363	$9,777	$9,787
Graduation Rate	82%	89%	85%	80%	91%	88%	93%	90%	87%
Average ACT Scores	16	16	14	15	16	15	14	14	14
% HS Graduates Attending College	6%	25%	4%	12%	20%	13%	6%	6%	8%
# of Failing Schools	0	0	2	1	1	1	1	2	1
Overall State Ranking Out of 67 Counties	24th	53rd	64th	55th	47th	50th	65th	61st	67th

Sources: Alabama State Department of Education, 2013-2014; Kids Count for Alabama Data Book, 2013; Alabama Connection

development, guided career exploration, interaction with STEM professionals, mentoring, and more, Project Engage complements STEM content delivered in the classroom and provides motivation for persistence in STEM fields. Improving the quality of STEM education in the region has the potential to develop highly qualified STEM professionals who can positively impact the STEM workforce and break the region's poverty cycle. The framework for Improving Undergraduate STEM Education proposed in this book supports the growth of the undeveloped talents of rural students. A nation-wide adoption of the framework has the capacity to increase the STEM workforce through capitalizing on the capabilities of an overlooked population.

CONCLUSION

Overall the U.S. lags behind its global counterparts in building a strong and extensive STEM workforce. There exists an untapped source of potential (i.e. rural students) for building the U. S. STEM workforce capacity. Issues such as lack of academic preparedness, lack of professional role models, lack of exposure to STEM professions, etc., which have historically impeded the capability of this demographic group must be addressed. Programs which provide enriched STEM experiences as well as career guidance, academic support, training in specific STEM skills, and overall general support are needed to both attract and retain students in STEM (Morgan & Gerber, 2016) fields, especially those from underrepresented groups such as rural and/or minority students. Project Engage is one such program that has the potential to provide the needed, scaffolded support to overcome the barriers which often impede the success of students, particularly those from rural America. Project Engage has shown success in providing a supportive environment that encourages student persistence in STEM disciplines. The subsequent chapters will provide a more detailed examination of Project Engage and the impact of the Project's components on undergraduate STEM students from rural areas, as well as the impact on K-12 high school students in the project service areas. The students majoring in STEM fields will be the STEM professionals of tomorrow, ultimately growing the U.S. STEM workforce. The challenge is overcoming the barriers that prevent the students from succeeding in STEM programs. A comprehensive review of each component of the project is presented as well as research findings of the impact of the project components. As the research unfolds, a four-dimensional framework for improving undergraduate STEM education of rural students evolves. Implementation of the Improving STEM Education Framework developed by the researchers stands to improve the persistence of rural students in STEM disciplines and therefore potentially increase the STEM workforce pool.

ACKNOWLEDGMENT

This research was supported by the U. S. Department of Education Minority Science and Engineering Program [grant number P120A120099] and the U. S. Department of Education Minority Science and Engineering Program [grant number P120A160122].

REFERENCES

Alabama School Connection. (2013). *The failing schools list—What does it really show us?* Retrieved from http://alabamaschoolconnection.org/2013/06/21/the-failing-schools-list-what-does-it-really-show-us/

Alabama State Department of Education. (2013-2014). Retrieved from https://www.alsde.edu/#

Allendoerfer, C., Wilson, D., Kim, M. J., & Burpee, E. (2014). Mapping beliefs about teaching to patterns of instruction within science, technology, engineering, and mathematics. *Teaching in Higher Education, 19*(7), 758–771. doi:10.1080/13562517.2014.901962

American College Testing Program, Inc. (2005). *Crisis at the core: Preparing all students for college and work.* ACT Inc.

American College Testing Program, Inc. (2006). *Reading between the lines: What the ACT reveals about college readiness in reading.* ACT Inc.

American Enterprise Institute. (2016, January). *The AP peak.* Washington, DC: Author.

Anderson, R., & Chang, B. (2011). Mathematics course-taking in rural high schools. *Journal of Research in Rural Education, 26*(1). Retrieved from http://jrre.psu.edu/articles/26-1.pdf

Aragon, S. (2016). *Teacher shortages: What we know.* Education Commission of the States. Retrieved from http://www.ecs.org/ec-content/uploads/Teacher-Shortages-What-We-Know.pdf

Avery, L. (2013). Rural science education: Valuing local knowledge. *Theory into Practice, 52*(1), 28–35. doi:10.1080/07351690.2013.743769

Babco, E. L. (2003). *Trends in African American and Native American participation in STEM higher education.* Commission on Professionals in Science and Technology. Retrieved from http://www.inpathways.net/STEM.pdf

Beeson, E., & Strange, M. (2000). Why rural matters: The need for every state to take action on rural education. *Journal of Research in Rural Education, 16*(2), 63–140.

Boynton, M. & Hossain, F. (2010). Improving engineering education outreach in rural counties through engineering risk analysis. *Journal of Professional Issues in Engineering Education and Practice, 136*(4), 224-232. doi: 10.1061(ASCE)EI.1943-5541.0000026

Business Roundtable. (2005). *Tapping America's potential: The education for innovation initiative*. Washington, DC: Business Roundtable.

Byun, S., Meece, J., & Irvin, M. (20121). Rural-nonrural disparities in postsecondary educational attainment revisited. *American Educational Research Journal, 49*(3), 412–437. doi:10.3102/0002831211416344 PMID:24285873

Carnegie Science Center. (2014). *Work to do: The role of STEM education in improving the Tri-State region's workforce*. Pittsburgh, PA: Campos Research Strategy. Retrieved from http://www.carnegiesciencecenter.org/csc_content/stemcenter/pdf/Work_to_Do_The_Role_of_STEM_Education_in_Improving_the_Tri-State_Regions_Workforce.pdf

Carsey Institute. (2009, Fall). *Students in rural schools have limited access to advanced mathematics courses* (Issue Brief No. 7). Durham, NH: Author.

Engberg, M. & Wolniak, G. C. (2013). College student pathways to the STEM disciplines. *Teachers College Record, 115*(1).

Fowler, J., Butler, J. S., Cowen, J. M., Streams, M. E., & Toma, E. F. (2013). Public employee quality in geographic context: A study of rural teachers. *American Review of Public Administration*, 1–19.

Gagnon, D., & Mattingly, M. J. (2012). *Beginning teachers are more common in rural, high-poverty, and racially diverse schools (Issue Brief No. 53).* Durham, NH: Carsey Institute.

Guckian, K., & Sarrio, J. (2012). Georgia schools lay unequal foundations for college. *Atlanta Journal-Constitution.* Retrieved from www.ajc.com

Haller, E. J., & Virkler, S. J. (1993). Another look at rural-nonrural differences in students' educational aspirations. *Journal of Research in Rural Education, 9*(3), 170–178.

Hossain, M. M. & Robinson, M. G. (2012). How to motivate US students to pursue STEM (Science, Technology, Engineering and Mathematics) careers. *US-China Education Review,* 442-45.

Ingels, S. J., Pratt, D. J., Rogers, J. E., Siegel, P. H., & Stutts, E. S. (2004). *Education longitudinal study of 2002: Base-year data file user's manual (NCES 2004-405). National Center for Education Statistics, Institute of Education Sciences.* Washington, DC: Department of Education.

Kids Count for Alabama Data Book. (2013). Retrieved from http://datacenter.kidscount.org/data#AL/2/0/char/0

Lee, L., & Summers. (2003). *Beyond the interstate: The crisis in rural Alabama.* Retrieved from www.auburn.edu/outreach/publications/cross_connections.pdf

Mader, J. (2014). Report: STEM education lacking in rural areas. *Education Week.* Retrieved from http://blogs.edweek.org/edweek/rural_education/2014/10/_report_stem_education_lacking_in_rural_areas.html

Marder, M., Brown, R. C., & Plisch, M. (2017). *Recruiting teachers in high-needs STEM fields: A survey of current majors and recent STEM graduates.* Retrieved from https://www.aps.org/policy/reports/popa-reports/upload/POPASTEMReport.pdf

Morgan, M. C., & Gerber, M. M. (2016). Utilizing factor analysis to inform the development of institutionally contrived experiences to increase STEM engagement. *Community College Journal of Research and Practice*, *40*(3), 204–218. doi:10.1080/10668926.2015.1112322

National Academy of Sciences. (2012). *A framework for K-12 science education: Practices, crosscutting concepts, and core ideas.* Washington, DC: The National Academies Press. Retrieved from http://nap.edu/13165

National Academy of Sciences, National Academy of Engineering, and Institute of Medicine. (2011). *Rising above the gathering storm revisited: Rapidly approaching category 5.* Condensed version. Washington, DC: The National Academies Press. Retrieved from http://nap.edu/12999

National Center for Education Statistics. (2013). *The status of rural education.* Retrieved from: https://nces.ed.gov/programs/coe/indicator_tla.asp#info

National Research Council. (2011). *Successful K-12 STEM education: Identifying effective approaches in science, Technology, Engineering and Mathematics.* Washington, DC: The National Academies of Press. Retrieved from http://nap.edu/13158

National Science Board. (2014). Science and Engineering Indicators 2014. Arlington, VA: National Science Foundation (NSB 14-01).

NGSS Lead States. (2018). *The need for standards*. Retrieved from https://www.nextgenscience.org/

Niu, L. (2017). Family socioeconomic status of a national sample. *College Student Journal*, *51*(2), 298–312.

Ossola, A. (Nov. 10, 2014). The challenge of teaching science in rural America. *The Atlantic*. Retrieved from http://www.theatlantic.com/education/archive/2014/11/the-challenge-of-being-a-rural-science-teacher/382309/

Pitchford, B. N. (2016). *The rural effect: How male and female students' STEM perceptions are influenced by attending a rural Georgia high school* (Doctoral dissertation). Retrieved from: https://vtext.valdosta.edu/xmlui/bitstream/handle/10428/2234/pitchford_dissertation_2016.pdf?sequence=1&isAllowed=y

Prescod, D. (2014). *The influence of a career planning STEM explorations course on vocational maturity, career decidedness, and career thoughts for undergraduate students* (Doctoral dissertation). Retrieved from http://stars.library.ucf.edu/cgi/viewcontent.cgi?article=4041&context=etd

President's Council of Advisors on Science and Technology. (2010). *Prepare and inspire: K-12 education in science, technology, engineering, and math (STEM) for America's future*. Retrieved from http://www.whitehouse.gov/sites/default/files/microsites/ostp/pcast-stem-ed-final.pdf

Provasnik, S., KewalRamani, A., Coleman, M. M., Gilbertson, L., Herring, W., & Xie, Q. (2007). *Status of education in rural America* (NCES 2007-040). Washington, DC: National Center for Educational Statistics, Institute of Education Sciences, U.S. Department of Education. Retrieved from http://nces.ed.gov/pubs2007/2007040.pdf

Raper, A. F. (1936). *Preface to peasantry: A tale of two Black Belt counties*. Columbia, SC: University of South Carolina Press.

Riegle-Crumb, C., & Grodsky, E. (2010). Racial ethnic differences at the intersection of math course-taking and achievement. *Sociology of Education*, *83*(3), 248–270. doi:10.1177/0038040710375689

Schafft, K. A. (2016). Rural education as rural development: Understanding the rural school-community well-being linkage in a 21st-century policy context. *Peabody Journal of Education*, *91*(2), 137–154. doi:10.1080/0161956X.2016.1151734

Schmidt, W. H. (2011). *STEM reform: Which way to go?* Paper presented at the National Research Council Workshop on Successful STEM Education in K-12 Schools. Retrieved from http://www7.nationalacademies.org/bose/STEM_Schools_Workshop_Paper_Schmidt.pdf

Spencer, K. (2017, January 22). *Not all towns are created equal, digitally: How a Colorado school district struggles to give its students a technology boost* (The Hechinger Report). New York, NY: Teachers College at Columbia University. Retrieved from http://hechingerreport.org/not-all-towns-arecreated-equal-digitally

Strange, M. (2011). Finding fairness for rural students. *Phi Delta Kappan*, *92*(6), 8–15. doi:10.1177/003172171109200603

Trust, T. (2017). *Why rural matters 2015-2016: Understanding the changing landscape*. Retrieved from https://www.instituteforchildsuccess.org/publication/rural-matters-2015-2016-understanding-changing-landscape/

U. S. Department of Education. (2017). *Minority Science and Engineering Program*. Retrieved from https://www2.ed.gov/programs/iduesmsi/index

U. S. Government Accountability Office. (2005). *Higher education: Federal science, technology, engineering, and mathematics programs and related trends*. Washington, DC: U. S. Government Accountability Office.

U.S. Department of Labor. (2007). *The STEM workforce challenge: The role of the public workforce system in a national solution for a competitive science, technology, engineering, and mathematics (STEM) workforce*. Retrieved from https://www.doleta.gov/youth_services/pdf/STEM_Report_4%2007.pdf

Versypt, J. J., & Versypt, A. N. (2013). Mapping rural students' STEM involvement: Case studies of chemical engineering undergraduate enrollment in the states of Illinois and Kansas. *120th ASEE Annual Conference and Exposition*.

Williams, D. T. (2010). *The rural solution: How community schools can reinvigorate rural education*. Retrieved from www.americanprogress.org/issues/2010/09/pdf/ruralschools.pdf

KEY TERMS AND DEFINITIONS

Black Belt: A region stretching from Texas to Virginia with rich fertile black soil. This region is characterized by high poverty, low levels of educational attainment, and few economic opportunities.

Minority Science and Engineering Improvement Program: A U. S. Department of Education grant program designed to improve STEM education at predominantly minority institutions of higher education and to increase the participation of underrepresented ethnic minorities, particularly minority women, in scientific and technological careers.

Minority Students: Refers to students who are American Indian, Alaskan Native, Black (not of Hispanic origin), Hispanic, and Pacific Islander. For this study, minority students are predominantly Black, not of Hispanic origin.

Project Engage: A comprehensive, capacity-building program which provides strategic mentoring, career exploration, as well as knowledge and skills to prepare undergraduate STEM majors, specifically minority students from rural areas for future STEM careers.

Rural Schools: Typically, small schools which are geographically isolated.

Rural Students: Students who attend rural schools. Often characterized by high poverty.

STEM Pipeline: The flow of students from college STEM majors to professional STEM careers.

Underfunded: Describes the low financial expenditure per student in rural schools.

Underrepresented Groups: Refers to populations who are not represented in STEM professions in proportions equal to White STEM workers. This term typically refers to minority groups, but may also refer to females.

Chapter 2
Project Engage:
Solutions for Recruiting, Retaining, and Reinforcing STEM Majors From Rural Areas

ABSTRACT

As the nation seeks to diversify the STEM workforce, looking toward the rural areas of the United States affords an opportunity to increase the flow of students into the STEM pipeline. However, rural students face many roadblocks on the way to STEM careers including poor educational attainment, lack of STEM experiences, few STEM role models, inadequate technology preparation, and low motivation to pursue STEM. This chapter explores those barriers and proposes Project Engage as an avenue for breaking down the barriers. The chapter identifies and explains the major components of Project Engage: strategic mentoring, STEM engagement activities, and guided career exploration. These combined components create a comprehensive framework for addressing the barriers imposed by the nature of growing up "rural." Statistical data is presented that reveals an increase in recruitment of STEM majors as well as positive results of retention.

DOI: 10.4018/978-1-5225-6341-9.ch002

INTRODUCTION

Students from rural areas are often from high poverty areas and face multiple challenges as they seek postsecondary education. In order to capitalize on the number of rural students who are currently an underrepresented group in STEM, efforts must be employed to dismantle the barriers and provide a supportive educational environment during the undergraduate years. Project Engage is a Minority Science and Engineering Program (MSEIP) grant project funded by the U. S. Department of Education supporting the recruitment and retention of underrepresented groups, specifically minorities, into STEM. Project Engage utilizes three general components: strategic mentoring, STEM engagement, and guided career exploration, to encourage persistence in STEM. When combined, these components support a framework for improving STEM education overall, which will be explained in its entirety in Chapter 6. However, the purpose of this chapter is to provide an overview of Project Engage and these three components as they relate to addressing the needs of rural students pursuing education in STEM disciplines.

PROJECT ENGAGE OVERVIEW

Project Engage is a program funded by the U.S. Department of Education MSEIP grant. This program assists “predominantly minority institutions in effecting long-range improvement in science and engineering education programs and increasing the flow of underrepresented ethnic minorities, particularly minority women, into science and engineering careers.” MSEIP provides four types of grants: 1) institutional project grants supporting the implementation of a comprehensive science improvement plan, which may include any combination of activities for improving the preparation of minority students for careers in science; 2) special project grants supporting activities that improve quality training in science and engineering at minority institutions or enhance the minority institutions’ general scientific research capabilities; 3) cooperative project grants assisting groups of nonprofit accredited colleges and universities to work together to conduct a science improvement program; and 4) design project grants assisting minority institutions that do not have their own appropriate resources or personnel to plan and develop long-range science improvement programs (U. S. Department of Education, 2017).

Project Engage

Project Engage was originally a three-year institutional project grant from 2012 to 2015 with a one-year extension. Investigators designed a comprehensive framework for improving STEM education in the rural Black Belt, the region surrounding the small, four-year university where most of the students call home. The framework contains four dimensions: *motivating* students, *experiencing* STEM, *engaging* in STEM, and *scaffolding* student learning. All components of Project Engage are designed to address one of the four dimensions of the framework in an effort to better prepare rural students to succeed in college and enter the STEM workforce. The activities and data results presented in this book evolve from the first grant award (2012-2015). However, a second award was made in 2016 (Project Engage 2.0) with additional components. With data from the first award, the second grant proposal was revised to incorporate needed changes including adding more of a focus on academic support (tutoring, supplemental instruction and a STEM resource room) as well as STEM Clubs (robotics, app design, and 3-D printing) to maintain the engagement of the STEM undergraduates until degree completion.

The Project Leadership Team

Research has pointed to the success of interdisciplinary faculty teams consisting of STEM and Education faculty in improving postsecondary STEM experiences (Bouwma-Gearhart, Perry & Presley, 2014). According to Ejiwale (2014),

Collaboration among scholars to form cluster groups in the university will enable further knowledge sharing with effective cost saving in the preparation of the future workforce for the emerging fields that will be developed by the integration of STEM disciplines. (p. 39)

With the collaboration among the faculty from the College of Natural Science and Mathematics, College of Education, and College of Business and Technology from a small, rural, four-year university, Project Engage implemented a comprehensive STEM improvement plan intended to increase the pipeline of underrepresented ethnic minorities, particularly minority women, into scientific, mathematical, engineering, and technological careers. The Project Team was purposefully assembled so as to be representative of all areas of STEM (Science, Technology, Engineering, Mathematics, and Education) in an effort to provide freshmen STEM majors with at least one faculty STEM mentor and role model.

What Constitutes STEM?

The Project Team, like most Americans, struggled with identifying which academic fields constituted STEM. It seems simple: Science, Technology, Engineering, and Mathematics, just like the acronym; however, one of the problems with STEM education is the lack of a clear definition. Not having a steadfast definition of STEM hinders efforts to improve STEM education. For some, STEM education refers to a program, curriculum, activity, etc., that combines at least two of the fields (Brown, 2012). The U. S. Department of Education (2007) declared, "Science, Technology, Engineering, and Mathematics education programs are defined as those primarily intended to provide support for, or to strengthen, science, technology, engineering, or mathematics (STEM) education at the elementary and secondary through postgraduate levels, including adult education" (p. 11). To clearly understand who Project Engage served and continues to serve, it is important to know the definition of STEM used. During the first grant award period (2012-2016) the Team adhered to a narrow definition of STEM including all of the STEM fields offered at the small rural institution including science fields (i.e. biology, biology education, chemistry, general science education, pre-professional sciences), computer science, mathematics, mathematics education, and engineering fields. For the subsequent award in 2016, the Team expanded its definition to include all of the fields promoted by the U. S. Department of Education's MSEIP definition. According to the U. S. Department of Education's MSEIP (2017) "The term 'science' means the biological, engineering, mathematical, physical, behavioral, and social sciences, and history and philosophy of science; also included are interdisciplinary fields which are comprised of overlapping areas among two or more sciences". The broader definition of "science" allowed the new Project Engage program (Project Engage 2.0) to increase the number of underrepresented women and minorities included in the program thus expanding the reach of the project.

PROJECT ENGAGE: ADDRESSING BARRIERS TO STEM EDUCATION

Low numbers of students entering post-secondary STEM fields coupled with high STEM attrition rates is cause for concern. Chen's study (2013) of a cohort of students entering either a bachelor's or associate's degree program in the 2003-2004 academic year revealed only 28% declared a STEM major over a six-year time period, and "a total of 48 percent of bachelor's degree students and 69 percent of associate's degree students who entered STEM fields between 2003 and 2009 had left these fields by spring 2009" (p. iv). Approximately half of the students changed to non-STEM majors, while the others left college altogether. Likewise, central to the need for Project Engage was a pattern of low science, math, and technology retention rates from enrollment to completion. A review of related research suggests positive relationships between early undergraduate student engagement and retention rates beyond the freshman year (Schneider, Bickel, & Morrison-Shetlar, 2015; Palmer, Maramba, & Elon, 2011). Therefore, Project Engage specifically focused on freshman students to achieve the purpose of increasing student achievement and retention rates of minority STEM majors.

Project Engage's four-dimensional comprehensive approach to improving STEM education outcomes was designed and implemented to address the gaps identified in rural students, including poor educational preparation, lack of STEM experiences, inadequate technology preparation, lack of motivation, and lack of STEM professional role models. By implementing a comprehensive plan to recruit, retain, and graduate students, chiefly underrepresented minority students from the rural Black Belt, Project Engage is tapping into a previously ignored group of potential STEM professionals. The Project Engage model has an overarching dimension of "Motivating". Motivating STEM undergraduate students plays a key role in retaining them in their STEM majors. Supporting the motivating dimension of the Improving STEM Education Framework are three additional dimensions: experiencing, engaging, and scaffolding. Collectively, these four dimensions of the Project Engage program improve the undergraduate STEM education experience. To achieve the four dimensions of the Improving STEM Education Framework, Project Engage used three overarching components: strategic mentoring, STEM engagement, and guided career exploration.

Poor Educational Preparation

Many rural students entering the university as freshmen are first-generation college students from rural areas of Alabama's Black Belt. First generation college students tend to be less academically prepared for the rigors of college-level coursework (Balemian & Feng, 2013; Choy, 2001) and often have lower grade point averages (GPAs), scores on college entrance exams, as well as critical thinking assessment scores (Katrevich & Aruguete, 2017). Access to educational resources which support student learning of STEM is also not equitable among racial, ethnic, and even geographically defined groups. White and Asian students typically have greater access to high schools with a wide range of advanced math and science courses compared to minority students. Across the nation, K-12 schools in select communities with the greatest resources have access to the best equipment and innovative STEM teaching practices, while those from economically disadvantaged areas are forced to "make do" with limited resources (U. S. Department of Education, 2016). This is true for K-12 education in schools across the Black Belt. Students entering post-secondary education as beginning freshmen often are not prepared due to inadequate educational backgrounds and, as a result, frequently drop out before completion of their freshman and/or sophomore year. For instance, in fall 2014 the university had 662 on-campus freshmen; 95 were enrolled in remedial English with 77% minority and 93 were enrolled in remedial math with 82% minority. Furthermore, a large portion of minority freshmen entering the institution (30%) have an ACT composite score of 18 or below, causing many to be admitted conditionally. Adding to the evidence illustrating the unpreparedness of local K-12 students for postsecondary education is the 13% high school dropout rate for the nine surrounding county school districts with only a small percentage of students from these districts graduating high school and attending college (on average 5.8%). The preponderance of evidence indicates students from this rural area who enter postsecondary education are lacking in the requisite knowledge and skills to be successful in a STEM major.

Lack of STEM Experiences

Freshmen from rural areas entering post-secondary education do not have the experience or preparation at secondary or elementary levels to select a STEM fields as a career choice. Niu (2017) reported that low socioeconomic students are usually not prepared to make informed decisions regarding STEM careers suited to their skill level, thus resulting in less success in post-secondary STEM education. The U. S. Department of Education's Office for Civil Rights found that rural, suburban, and urban high schools lack needed advanced math and science course offerings to fully prepare students for STEM higher education. Only 53% of American schools offer calculus while 63% offer physics, both subjects vital for preparing secondary students for post-secondary success in a STEM field. Access is even lower in high schools with predominantly minority student populations. Students often do not have positive or realistic outlooks for a career in STEM due to the educational, geographical, and economical environments of the schools they attended. Stipanovich and Woo (2017) concluded that "environmental factors, such as counseling, family expectations, and available resources, played a substantial role in influencing the students' interests, experiences, and perceptions in their pursuits of STEM careers" (p.192). Students need to be prepared, motivated, and counseled to choose STEM careers early to assure adequate preparation and success. Career exploration including opportunities to receive and investigate information on STEM careers, partake in self-interest career inventories, and participate in career planning promotes confidence in career choices (Tang, Pan, & Newmeyer, 2008; Turner & Lapan, 2002) and is especially important for ethnic minorities (Stipanovich & Woo, 2017). Career exploration is a key component in Project Engage. Providing students with career exploration opportunities offers exposure to STEM role models and increases students' abilities to make informed decisions regarding STEM career choices.

Inadequate Technology Preparation

Across America, limited access to broadband internet still impedes access to educational resources such as online courses, electronic research, streaming videos, and more. Approximately 26 million citizens lack internet access at home with 70% of those located in rural areas (Hassel & Dean, 2015). In 2014, Digital Learning Now rated Alabama public schools a D (overall 60%) based on the 10 elements of High-Quality Digital Learning. Specifically noted by the digital learning report card were low scores in the areas of access to digital content and online courses (63%), high quality digital instruction (64%), funding (25%), and infrastructure that supports digital learning (63%). Two-thirds of the schools across the nation have broadband speeds of 25 Mbps or less that limits access to educational resources such as streaming videos and virtual labs which require high speed internet access (Hassel & Dean, 2015). Lack of available technology impedes students' ability to successfully complete technology projects, internet research or online classes both in and out of school. Many entering students from the rural Alabama Black Belt have not been exposed to technology as a teaching/learning tool. Students come with little knowledge or experience in the use of technology and up-to-date pedagogy; although the schools in their communities may possess technology, they often lack high speed broadband internet access to make adequate use of it. To exacerbate the problem, according to a 2015 report from the Federal Communications Commission, 56% of the population in rural Alabama lacks internet access at home highlighting the digital divide between rural and urban Alabamians. To fill this gap, Project Engage provides activities exposing students to the breadth and depth of opportunities in STEM fields through use of technology. Participants in Project Engage were exposed to hands-on technology activities, using technology as tools in solving real world problems, and STEM careers that rely heavily on technology.

Lack of Motivation

Despite a lack of educational preparedness and resources, many students from rural areas recognize STEM careers as a means to improve quality of life and are motivated to enter a STEM field. Student motivation to pursue a STEM career is influenced by a variety of factors including self-concept of ability in math and science, task values in math and science, altruistic values, family values, and monetary values (Wang, Ye, & Degol, 2016). Students from rural areas entering college, often as first-generation college students, lack the intrinsic motivation and self- efficacy to be successful in STEM majors. Motivational events and activities are needed to help students realize their full potential. Pedagogical practices steeped in inquiry based learning that promote critical thinking and provide authentic learning experiences increase student interest and motivation relative to STEM (McDonald, 2016). Intervention programs that provide academic and motivational support have shown positive results on student success such as higher GPAs and higher retention rates from freshman to sophomore years (Kassaee & Rowell, 2016). Students targeted by Project Engage benefited from STEM focused activities to increase self-efficacy and motivation for success in STEM. Project activities are intentionally designed based on a comprehensive framework for improving STEM education which is detailed in the final chapter. However, it should be revealed here that the heart of the framework is motivation. The other three components (i.e. experiencing, engaging, and scaffolding) provide the building blocks that supports the ultimate motivating purpose.

Lack of Professional Role Models

Nationally, women and minorities are underrepresented in STEM professions. The Black Belt Region lacks a network of STEM professionals, particularly women and minority role models, in all STEM fields. Additionally, extreme shortages in professional services and in math and science educators exist. Exposure to STEM role models positively affects students' (STEM and non-STEM undergraduates) sense of belonging and interest in STEM. Thus, exposure to STEM role models is another method for recruiting students into STEM majors and ultimately increasing the STEM workforce numbers (Rosenthal, Levy, London, Lobel, & Bazile, 2013; Shin, Levy & London, 2016). Project Engage furnished undergraduate STEM students with role models in the form of mentors, STEM guest speakers, and STEM professionals in their work environments. Furthermore, actively engaging students in a motivational program such as Project Engage, helps minority and high-need students from rural areas succeed in STEM disciplines and become role models for their communities, thus enhancing the future of the STEM professionals' pipeline in the Black Belt and beyond.

PROJECT ENGAGE ACTIVITIES

Project Engage activities were designed and implemented to address the aforementioned gaps so as to improve retention rate of STEM majors and, in particular, minority female STEM majors. These activities fall into three categories: strategic mentoring, STEM engagement, and guided career exploration. Together these categories have evolved over time into a four dimensional framework (motivating, experiencing, engaging, and scaffolding) for improving STEM education and producing accomplished STEM graduates. Each component plays a vital role in supporting the success of undergraduate STEM majors. As an introduction to Project Engage's activities, the following provides a summary of each component which will be examined in greater detail in subsequent chapters.

Strategic Mentoring

Mentors play a key role in nurturing the development of STEM students. Structured mentoring programs for undergraduate STEM students promote success among African American students (Griffin, Perez, Holmes & Mayo, 2010; Kendricks, Nedunuri, & Arment, 2013). Other researchers have had like success in positively impacting academic performance (Chester, Burton, Xenos, & Elgar, 2013) as well as social integration of at-risk students into the college community life (Morales, Ambrose-Roman, & Perez-Maldonado, 2016). Mentoring offers a three-fold usefulness for freshmen STEM majors: 1) STEM program knowledge, 2) STEM role models, and 3) a peer relationship. Strategic mentoring activities were accomplished through intentional mentoring activities performed within a STEM freshmen seminar class and targeted mentoring activities performed outside of the seminar class. Through the STEM freshmen seminar class, students were paired with a STEM upperclassmen mentor of a like major, introduced to other STEM faculty, participated in early STEM engagement activities under the guidance of STEM faculty, and received career exploration support from a graduate student career counselor. Outside class, Project Engage freshmen students met individually with a mentor once per week during the fall and met monthly with a small cohort of fellow mentees facilitated by a Project Engage student mentor.

STEM Engagement Activities

Providing opportunities for students to experience or "engage" in STEM is another aspect of Project Engage. Engaging in STEM not only in content classrooms, but outside the classroom as well, is essential for developing STEM self-efficacy and maintaining motivation. Project Engage used competitive STEM events to engage undergraduate STEM majors in varied STEM activities. One of the engagement activities was an Invention Convention. It is a competitively judged event in which individual undergraduate STEM students and regional high school students showcased their STEM-related inventions and discoveries. Several twists on the Invention Convention have included designing, building and testing a boomerang, a space rover, and an egg drop carrier. The level of engagement required for program participants to take part in a competition of this nature required them to be actively engaged in the project. Engaged students were more motivated, thereby increasing the rate of retention in a STEM degree program. A second engagement activity was the STEM Challenge. It was a competitive, day-long team event where STEM college students and regional high school students competed in teams of five students in four major events to tackle science, technology, engineering, and mathematic dilemmas using tools and resources provided by Project Engage. In addition to being an engagement event for program participants and an outreach effort for area schools and the community at large, this event also served as an opportunity for Project Engage students to hone organizational, leadership, and public relations skills through their involvement in the planning process. Finally, robotics competitions have actively engaged the more math and engineering inclined college students. Learning to design, build, and program a robot to maneuver an obstacle course or complete a task provided opportunities for gaining intricate knowledge and skills related to engineering design and programming while simultaneously learning to work as part of team. Activities that inspire students to use their knowledge and skills to solve problems or meet project design requirements are essential. Way too often, content is taught in isolation from application. The students who are needed to fill the STEM pipeline are those who can apply their content knowledge and skills to solve real problems outside the classroom.

Guided Career Exploration Activities

In addition to strategic mentoring and providing engaging STEM opportunities, guided career exploration activities are crucial for supporting the persistence of undergraduates from freshmen year to graduation. The Black Belt area is characterized by a predominantly low income, minority population. Research has revealed that Black students are more likely to choose a STEM major in college if other factors such as SES are comparable. Likewise, the gender gap in STEM enrollment decreases as SES increases (Niu, 2017). Informed by research in the literature, Project Engage attempts to compensate for the low SES conditions with concentrated career guidance efforts. Encouraging minority students to pursue scientific and technological careers is a primary focus of Project Engage and is critical to increasing the flow of underrepresented populations in the STEM community. Guided career exploration for Project Engage participants was accomplished through participation in career exploration field trips, career counseling services, increased access to STEM professionals, and focused STEM career exploration coursework. Collaborations between Project Engage faculty and STEM colleagues at other educational institutions and businesses open doors for future placement of students in internships and cooperative education programs. Each year, two career exploration field trips and one leadership/team building field trip were organized. Such experiences allowed students to tour STEM-related industries and facilities otherwise unavailable to them, personally interact with professionals currently working in STEM-related disciplines, and develop highly needed leadership and team building skills. Project Engage graduate counselors worked in conjunction with science, math, and technology faculty to invite experts in STEM disciplines to consult with students via personal visit, conference call, webinar, and/or SKYPE. With access to STEM professionals, Project Engage students were encouraged to ask questions about professional opportunities and career outlooks, as well as inquire about specific research topics of personal or shared interest. Through this endeavor, Project Engage participants regularly interacted with real-life practitioners working in STEM-related fields, resulting in a greater awareness of working environments, thereby increasing student interest in pursuing STEM careers. Furthermore, STEM students were provided opportunities for extensive research into STEM careers as wells as opportunities to explore their own strengths, weaknesses, and propensities for various careers. Two courses, the freshmen STEM seminar course and the STEM career exploration course, afforded undergraduate STEM students ample occasion to investigate STEM employment opportunities. By providing a strong career exploration focus, students were exposed not only to the careers available in STEM, but also to the knowledge, skills and dispositions needed for these careers. Choosing a STEM major/career that meets the interests of the students increases the chances of retention.

SUMMARY AND FUTURE RESEARCH DIRECTIONS

Project Engage was designed to increase both the number of students enrolled and retained in a STEM major. Prior to Project Engage, STEM enrollments at the institution were less than desired. For instance, in 2006, prior to the funding of Project Engage, 232 freshmen declared either mathematics or science as a major. Looking at the data from four years later (2010), only 27 students (11.6%) completed a degree in science or mathematics. The implementation of Project Engage intended to increase retention rates of STEM majors. Project Engage's initial cohort from 2012 (N=60) had a retention rate of 23% over four years (Figure 1). Subsequent years hold promise for an even higher four-year retention rate.

Figure 1. Project Engage participant retention rates in 2016

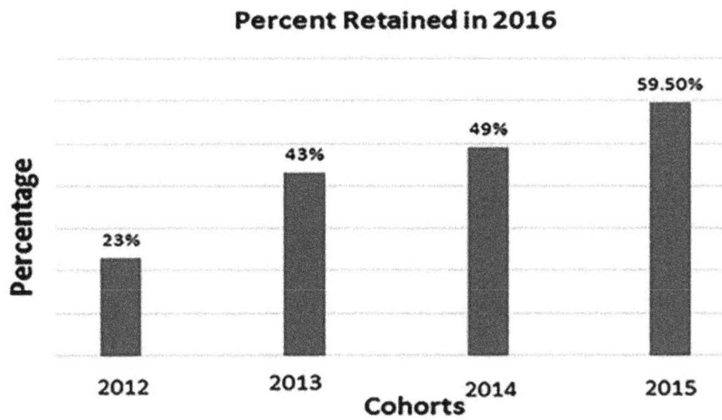

	n (Original)	n at the End of 2016	% Retained	Number of Years
2012 Cohort	60	14	23%	4
2013 Cohort	61	26	43%	3
2014 Cohort	82	40	49%	2
2015 Cohort	84	50	59.50%	1

The first four years of Project Engage have proven success in contributing to recruitment and retention of STEM undergraduate majors. From 2012 to 2016, Project Engage enrolled 287 freshmen STEM majors with 60 in 2012-2013, 61 in 2013-2014, 82 in 2014-2015, and 84 in 2015-2016. The retention rates after one year of enrollment at Project Engage are shown in Figure 2. The Project Engage program has continually been revised based on data collected each year after the initial implementation in 2012. As illustrated in Figure 2, the first year retention rates for cohorts 2013, 2014, and 2015, when averaged are approximately 20% higher than the 2012 cohort.

The strategies implemented by Project Engage are specifically addressing the barriers to STEM careers of students in the rural Alabama Black Belt. Other rural areas may have barriers to STEM careers that differ, which may limit the generalization of the strategies proposed here. However, results from the retention data for freshman to sophomore year are positive. Future research is needed to examine new methods for retaining students beyond the sophomore year to graduation. The new grant award for Project Engage includes strategies such as STEM clubs to affect the outcomes for sophomore, junior, and senior students. Research exploring the success of the new strategies is needed to assess their impact on retention.

Figure 2. Project Engage cohort enrollment and retention numbers after one year

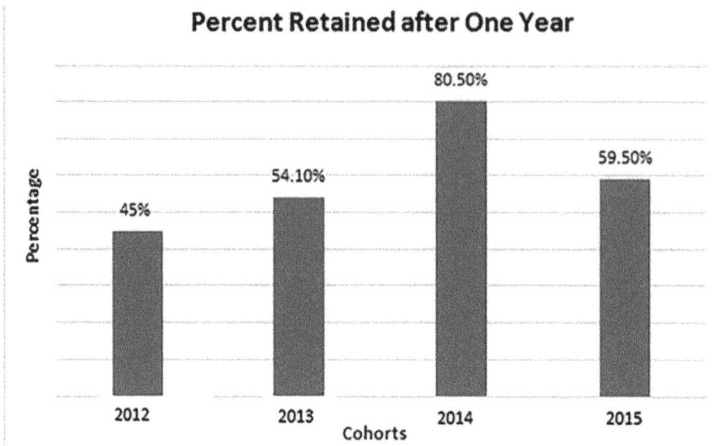

	n (Original)	n (After 1 Year)	Percent Retained
2012 Cohort	60	27	45%
2013 Cohort	61	33	54.10%
2014 Cohort	82	66	80.50%
2015 Cohort	84	50	59.50%

CONCLUSION

The overall goal of the MSEIP grant is to attract and retain underrepresented minority groups in STEM. Project Engage was developed as a comprehensive approach to recruiting STEM students from the rural Alabama Black Belt area, addressing the needs of undergraduate STEM students, and retaining them through graduation. Results from enrollment trends in STEM at this small rural four-year institution have shown an increased STEM enrollment supporting the efforts of Project Engage. Similarly, the retention rate of STEM students in Project Engage have shown a positive increase over the baseline data prior to implementation. In the next few chapters, the main components of Project Engage: strategic mentoring, preparing future STEM professionals, and guided career exploration will be explored and results from evaluation of the activities will be revealed.

ACKNOWLEDGMENT

This research was supported by the U. S. Department of Education Minority Science and Engineering Program [grant number P120A120099] and the U. S. Department of Education Minority Science and Engineering Program [grant number P120A160122].

Project Engage could not have been implemented without the dedication of STEM faculty members, Dr. Balakrishna Gokaraju, Dr. Frank Rogers, and Dr. Ketia Shumaker.

REFERENCES

Balemian, K., Feng, J., & College, B. (2013). *First generation students: College aspirations, preparedness and challenges*. Paper presented at the meeting of the Advanced Placement Annual Conference, Las Vegas, NV.

Bouwma-Gearhart, J. L., Perry, K. H., & Presley, J. B. (2014). Improving postsecondary STEM education: Strategies for successful interdisciplinary collaborations and brokering engagement with education research and theory. *Journal of College Science Teaching*, *44*(1), 40–47. doi:10.2505/4/jcst14_044_01_40

Brown, J. (2012). The current status of STEM education research. *Journal of STEM Education: Innovations & Research*, *13*(5), 7–11.

Chen, X. (2013). *STEM attrition: College students' paths into and out of STEM fields. Statistical analysis report* (Report No. 2014-001). Washington, DC: National Center for Education Statistics.

Chester, A., Burton, L. J., Xenos, S., & Elgar, K. (2013). Peer mentoring: Supporting successful transition for first year undergraduate psychology students. *Australian Journal of Psychology*, *65*(1), 30–37. doi:10.1111/ajpy.12006

Choy, S. (2001). *Students whose parents did not go to college: Postsecondary access, persistence, and attainment*. Washington, DC: National Center for Education Statistics.

Digital Learning Now. (2014). *Digital Learning Report Card.* Retrieved from http://www.digitallearningnow.com/report-card/state/2014/alabama-3/

Griffin, K. A., Pérez, D. II, Holmes, A. E., & Mayo, C. P. (2010). Investing in the future: The importance of faculty mentoring in the development of students of color in STEM. *New Directions for Institutional Research, 148*(Winter), 95–103. doi:10.1002/ir.365

Hassel, B. C., & Dean, S. (2015). *Technology and rural education*. Retrieved from http://www.rociidaho.org/wp-ontent/uploads/2015/03/ROCI_2015_RuralTech_Final.pdf

Kassaee, A. M., & Rowell, G. H. (2016). Motivationally-Informed interventions for at-risk STEM students. *Journal of STEM Education: Innovations and Research, 17*(3), 77–84.

Katrevich, A. V., & Aruguete, M. S. (2017). Recognizing challenges and predicting success in first-generation university students. *Journal of STEM Education: Innovations & Research, 18*(2), 40–44.

Kendricks, K. D., Nedunuri, K. V., & Arment, A. R. (2013). Minority student perceptions of the impact of mentoring to enhance academic performance in STEM disciplines. *Journal of STEM Education: Innovations & Research, 14*(2), 38–46.

McDonald, C. V. (2016). STEM education: A review of the contribution of the disciplines of science, technology, engineering and mathematics. *Science Education International, 27*(4), 530–569.

Morales, E., Ambrose-Roman, S., & Perez-Maldonado, R. (2016). Transmitting success: Comprehensive peer mentoring for at-risk students in developmental math. *Innovative Higher Education, 41*(2), 121–135. doi:10.100710755-015-9335-6

Niu, L. (2017). Family socioeconomic status of a national sample. *College Student Journal, 51*(2), 298–312.

Rosenthal, L., Levy, S., London, B., Lobel, M., & Bazile, C. (2013). In pursuit of the MD: The impact of role models, identity compatibility, and belonging among undergraduate women. *Sex Roles, 68*(7-8), 464–473. doi:10.100711199-012-0257-9 PMID:24497671

Schneider, B., Judy, J., & Mazuca, C. (2012). Boosting STEM interest in high school. *Phi Delta Kappan, 94*(1), 62–65. doi:10.1177/003172171209400112

Schneider, K. R., Bickel, A., & Morrison-Shetlar, A. M. (2015). Planning and implementing a comprehensive student-centered research program for first-year STEM undergraduates. *Journal of College Science Teaching*, *44*(3), 37–43. doi:10.2505/4/jcst15_044_03_37

Shin, J. L., Levy, S. R., & London, B. (2016). Effects of role model exposure on STEM and non-STEM student engagement. *Journal of Applied Social Psychology*, *7*(7), 410–427. doi:10.1111/jasp.12371

Stipanovic, N., & Woo, H. (2017). Understanding African American students' experiences in STEM education: An ecological systems approach. *Career Development Quarterly*, *65*(3), 192–206. doi:10.1002/cdq.12092

Tang, M., Pan, W., & Newmeyer, M. D. (2008). Factors influencing high school students' career aspirations. *Professional School Counseling*, *11*(5), 285–295. doi:10.5330/PSC.n.2010-11.285

Turner, S., & Lapan, R. T. (2002). Career self-efficacy and perceptions of parent support in adolescent career development. *The Career Development Quarterly*, *51*(1), 44–55. doi:10.1002/j.2161-0045.2002.tb00591.x

U. S. Department of Education. (2016). *STEM 2026: A vision for innovation in STEM education* (Rep.). Retrieved from https://innovation.ed.gov/files/2016/09/AIR-%20STEM2026_Report_2016.pdf

U.S. Department of Education. (2007). *Report of the academic competitiveness council.* Author.

U.S. Department of Education. (2017). *Minority Science and Engineering Improvement Program.* Retrieved from https://www2.ed.gov/programs/iduesmsi/faq.html

U.S. Department of Education, Office for Civil Rights. (2014). *Civil rights data collection: Data snapshot: College and career readiness* (Issue Brief No. 3). Washington, DC: Author. Retrieved from http://www2.ed.gov/about/offices/list/ocr/docs/crdc-college-and-career-readiness-snapshot.pdf

Wang, M., Ye, F., & Degol, J. L. (2017). Who chooses STEM careers? Using a relative cognitive strength and interest model to predict careers in science, technology, engineering, and mathematics. *Journal of Youth and Adolescence*, *46*(8), 1805–1820. doi:10.100710964-016-0618-8 PMID:27975183

KEY TERMS AND DEFINITIONS

Educational Preparation: Knowledge and skills acquired by K-12 students that equips them for a college career in STEM.

Guided Career Exploration: Activities that inform undergraduate STEM majors of their career choices in STEM.

Minority Science and Engineering Improvement Program: A U. S. Department of Education grant program designed to improve STEM education at predominantly minority institutions of higher education and to increase the participation of underrepresented ethnic minorities, particularly minority women, in scientific and technological careers.

Motivation: The desire to pursue an undergraduate degree in STEM or a STEM profession.

Project Engage: A comprehensive, capacity-building program that provides strategic mentoring, career exploration, as well as knowledge and skills to prepare undergraduate STEM majors, specifically minority students from rural areas for future STEM careers.

STEM Engagement: The act of being emotionally or physically involved with a STEM activity.

STEM Experiences: Knowledge and skills that you gain through the act of participating in STEM-related activities.

STEM Role Models: Professionals in the STEM workforce who K-12 and undergraduate students can aspire to be like.

Strategic Mentoring: A comprehensive plan for providing academic, social, and professional guidance to undergraduate college students.

Chapter 3
Scaffolding Undergraduate STEM Majors:
A Strategic Mentoring Program

ABSTRACT

Project Engage utilizes a scaffolded approach to strategic mentoring grounded in the social constructivist theory of Vygotsky. Peer mentors, a career counselor, and STEM faculty serve as three scaffolding layers of more knowledgeable others (MKOs) who are responsible for assisting the mentee in achieving his/her zone of proximal development (ZPD), a higher level of learning or understanding than could be achieved alone. Detailed information is shared in this chapter on the selection, training, and responsibilities of the peer mentors, given that they serve as the first level of scaffolding (i.e., primary mentors for the freshmen). The career counselor and Engage faculty members constitute levels two and three of scaffolding. A survey was administered to evaluate the effectiveness of the mentoring program. Results from the survey of the mentees revealed positive perceptions of the mentoring program.

INTRODUCTION

A structured mentoring program provides needed academic and social support, as rural students transition from secondary to post-secondary schooling. In this chapter, the authors present a three-tiered strategic mentoring process for scaffolding the learning and development of freshmen STEM students. A

DOI: 10.4018/978-1-5225-6341-9.ch003

mentoring strategy consisting of peer mentors, a career counselor, and STEM faculty creates a hierarchical community of mentors (MKOs) who guide students through their first year as college freshmen majoring in a STEM field. In this chapter, peer mentor selection, training, and responsibilities are described. Along with the description of the peer mentor, attention is given to the career counselor and STEM faculty mentoring roles. Finally, the authors present data from a mentoring survey administered to the freshmen mentees in order to gain their perspectives on the overall effectiveness of the program, academic benefits, persistence in the STEM major, and the likelihood of pursuing a STEM profession. The chapter's purpose is to recommend a mentoring program for improving the freshman year experience of STEM majors.

SCAFFOLDING FROM A CONSTRUCTIVIST VIEW

Social constructivism is a branch of cognitive constructivism emphasizing the social collaborative nature of learning. Unlike cognitivists such as Piaget who view knowledge as assimilated by learners through interacting and making sense of external stimuli, social constructivists see knowledge as constructed actively by learners through interaction and negotiation among people under a social cultural environment (Rogoff, 1990; Vygotsky, 1978). Social constructivism theory was developed by Vygotsky. His theories stress the fundamental role of social interaction in the development of cognition (Vygotsky, 1978; Wertsch, 1985). Contradicting Piaget's view of universal stages and content of cognitive development (i.e., sensorimotor, preoperational, concrete operational, and formal operational period) (Piaget, 1952; 1959), Vygotsky's cognitive development theory conceptualizes human cognitive structure as essentially socially constructed. Knowledge is therefore not simply constructed but co-constructed.

From a social constructivism point of view, learning is more than the assimilation of new knowledge. It is a collaborative process in which learners are integrated into a knowledge community of mutual thinking, problem solving, and decision making. Social constructivists see motivation as both intrinsic and extrinsic and believe that learners receive extrinsic motivation through the social learning process in the knowledge community. Social constructivism underpins learning environments that facilitate a community of learners encouraging interaction, discourse, and thoughtful reflection. In constructivist learning environments like this, scaffolding is an essential element providing "temporary frameworks to support learning and student performance beyond the learners' capacities" (Jonassen, 1999, p. 235). Scaffolding, as a teaching strategy, is theoretically related to two important principles of Vygotsky: The More Knowledgeable Other (MKO) and the Zone of Proximal Development (ZPD). These two principles are essential for a better understanding of Vygotsky's theory of social cognitive development and how it is related to scaffolding.

More Knowledgeable Other (MKO)

Vygotsky's (1978) social cognitive development theory gives recognition to the existence of the More Knowledgeable Other (MKO). The MKO is someone or something having a better understanding or a higher ability level regarding a particular task, process, or concept. The MKO does not necessarily have to be a person. It can be electronic tutors or other electronic performance support systems to facilitate and guide learners through the learning process. The key to MKOs is that they must have (or be programmed with) more knowledge about the topic being learned than the learner (Galloway, 2001).

Zone of Proximal Development (ZPD)

Vygotsky (1978, p. 85) distinguished between two developmental levels:

The level of actual development is the level of development that the learner has already reached, and is the level at which the learner is capable of solving problems independently. The level of potential development (the "zone of proximal development") is the level of development that the learner is capable of reaching under the guidance of teachers or in collaboration with peers. The learner is capable of solving problems and understanding material at this level that they are not capable of solving or understanding at their level of actual development; the level of potential development is the level at which learning takes place. It comprises cognitive structures that are still in the process of maturing, but which can only mature under the guidance of or in collaboration with others.

The ZPD is the area where a learner can reach with guidance and encouragement from MKOs and is beyond what a learner can achieve independently.

Scaffolding

Jerome Bruner was the American psychologist who first introduced the scaffolding theory in the late 1950s. Bruner's scaffolding theory is part of social constructivism theory and was particularly influenced by Vygotsky's work. The MKO and the ZPD are essential scaffolding components. Scaffolding is intended to enable learners to accomplish a learning task that they would not have been able to manage on their own and to bring them to a state of competence of being able to eventually complete such a task on their own (Wells, 1999). In terms of the MKO and the ZPD, scaffolding can be defined as the structured support from the MKO to the learners that allows them to reach the ZPD. Such structured support is reminiscent of the scaffolding used in the construction of a building. The same way as the scaffolding is gradually dismantled when the construction is completed, the support to learners can be gradually removed when they become more independent in their thinking and acquiring new knowledge or skills.

SCAFFOLDING MODEL OF PROJECT ENGAGE: STRATEGIC MENTORING

Project Engage employs a strategic mentoring strategy to address the gaps of poor educational preparation and lack of motivation that are characteristic of STEM majors upon entrance to college. The scaffolding theory of social constructivism was adopted in designing the strategic mentoring activities of Project Engage for the following reasons:

1. Endeavors undertaken by STEM professionals in our modern society are mostly a social process involving collaboration and team work. Project Engage students benefit by engaging in a community of learners and mentors to learn through social interactions, reflections, and collaboration. Likewise, all undergraduate students majoring in a STEM field benefit by learning to work collaboratively as members of a team.
2. Scaffolding can not only help Project Engage students with learning tasks they would not have been able to handle on their own, but also, enable them to develop the competence of working as STEM professionals after graduation.

3. A constructivist learning environment with scaffolding and social interactions provides Project Engage students with extrinsic motivations and allows them to persist as STEM majors and STEM professionals after graduation. Having a hierarchical and scaffolded mentoring program allows the freshmen undergraduates an opportunity to learn from multi-tiers of MKO's.

The following scaffolding model (Figure 1) was used in guiding the design and implementation of the strategic mentoring activities.

Junior, senior, or graduate level UWA STEM students serve as mentors for STEM undergraduate students enrolled in Project Engage. The mentors, together with faculty members teaching the freshmen seminar course (STEM 101), play the role of MKO to provide scaffoldings to Project Engage students through individual mentoring sessions, individual tutoring sessions, and UWA 101 sessions. Such scaffoldings were intended to lead UWA Project Engage students into their ZPD in expectation of increased persistence in STEM disciplines, improved STEM academic performance, and enhanced learning motivation for STEM.

Project Engage Strategic Mentoring

Mentoring is fundamental to the overall success of undergraduate STEM majors and specifically those from deprived environments. Students from rural backgrounds, due to geographic isolation and lack of exposure to STEM professionals, do not always have realistic perceptions of college life, the demands of a rigorous STEM major, nor the possibilities of STEM careers. Mentoring from the freshman year and continuing throughout the undergraduate career shows promise for increasing academic success (Chester, Burton, Xenos, & Elgar, 2013; Kendricks, Nedunuri, & Arment, 2013) as well as for promoting social and career guidance (Booker & Brevard, 2017). Moreover, mentoring also positively impacts the development of professional

Figure 1. Scaffolding model of Project Engage strategic mentoring

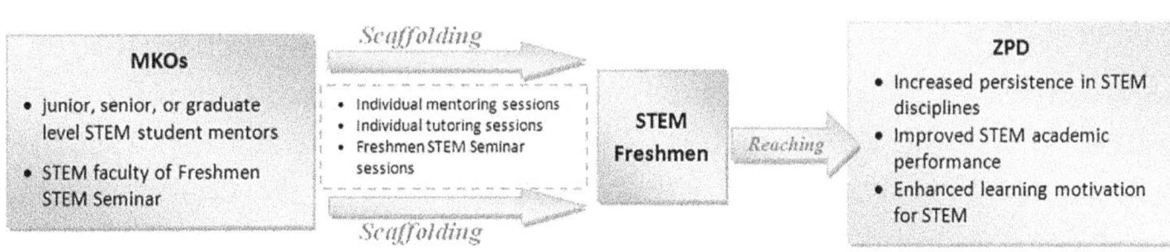

identity (Holland, Major, & Orvis, 2012), which may be lacking in STEM undergraduate students from rural areas. The scaffolded mentoring component of Project Engage is three tiered: mentoring by upperclassmen, mentoring by the career counselor, and mentoring by the faculty team (professors from education, science, engineering, mathematics, and technology disciplines). The three tiered approach assures students receive support from multiple perspectives needed for success in college and particularly in STEM. Project Engage's scaffolding supports a model of mentoring that combines hierarchal and community models to provide undergraduate minority STEM students needed support to be successful in a STEM major. Kobulnicky and Dale (2016) posited community mentoring models are more effective than traditional one-to-one mentoring. The community model employed by Kobulnicky and Dale, paired undergraduate STEM students with multiple mentors including peers, graduate students, and multiple faculty from STEM areas. Undergraduates benefited from the collaborative nature of participating in a community of STEM scholars. Likewise, the hierarchical mentoring approach used by Wilson et. al, paired undergraduate STEM students with STEM peers, staff, and faculty. Hierarchical mentoring improved retention rates of STEM undergraduates by positively impacting academic performance and self-identity as a STEM scholar (Wilson et al., 2012). Project Engage's three-tiered model incorporates both community and hierarchical mentoring concepts to accomplish a scaffolding for success.

The Engage Freshmen Cohort

Relying on social constructivism theory, Project Engage initially sought to establish a community of STEM support whereas mentors could lead mentees to their zone of proximal development. Project Engage took advantage of the freshmen STEM seminar course already established as a freshmen course requirement. Originally, freshmen seminar was a generic introduction to campus life including campus offices, academic services, financial aid, recreational activities, and such. At the onset of Project Engage, the first undertaking by the Project Team was to revamp the curriculum of the freshmen seminar class to include the proposed activities of Project Engage including STEM engagement, professional exploration field trips, STEM speakers, competitive STEM events, and mentoring by STEM upperclassmen, the career counselor, and the STEM faculty. The result of the revisions was a specific seminar course, STEM 101. A secondary outcome of the course was the creation of a cohort of STEM freshmen. The STEM freshmen cohort became part of a larger STEM community allowing them to work and study together as well as receive mentoring in the hierarchical structure of Project Engage.

Mentoring Tiers

Tier one of the scaffolded mentoring process encompasses mentoring by peers from like STEM majors. Mentoring by upperclassmen STEM majors is a crucial component of Project Engage, as it gives the underclassmen an experienced STEM role model to confer with on academic and/or social issues. Research has shown that peer mentoring supports students' commitment to and satisfaction with their STEM majors (Holland et al., 2012) as well as their overall STEM self-efficacy (Dennehy & Dasgupta, 2017). By grouping students with mentors who have the same STEM majors, the program establishes a foundation for building a personal relationship between mentor and mentee. The mentors have actual experience with the STEM faculty, courses, and general university experiences. Mentoring by an upperclassman with similar educational experiences provides the freshmen mentee a source of first-hand information for success in the chosen STEM field.

The second tier of the mentoring program is a provided by a graduate level career counselor responsible for guiding and supervising the upperclassmen mentors, as well as ongoing mentoring of the freshmen student participants as they progress to sophomores, juniors, and seniors. The career counselor serves as a bridge between mentees, upperclassmen mentors, and faculty mentors. The graduate student career counselor is an important intermediary between the peer and faculty mentors giving the mentees access to an older more experienced mentor, but one who may be perceived as more relatable than a faculty member. The career counselor, in conjunction with the Project faculty, leads the training of the mentors. Additionally, the career counselor actively guides freshmen participants as they progress forward in their STEM programs through meetings at least once per semester. Also, as the label "career counselor" indicates, the counselor oversees the development of professional career portfolios for the Project Engage participants.

The last tier of the Engage mentoring program is mentorship by STEM faculty. Previous research indicated collaboration among STEM faculty from the representative disciplines improved STEM education and was needed for successful research efforts (Bouwma-Gearhart & Bess, 2012; Bouwma-Gearhart, Perry, & Presley, 2014). The Project Engage faculty team is a collaborative assemblage of science, technology, engineering, mathematics, and education faculty. Representation from all disciplines provides discipline specific academic support and role models for STEM freshmen. Mentoring by STEM faculty positively affects motivation, academic performance, and persistence of minority STEM students (Griffin, Pérez, Holmes, & Mayo, 2010). The role of STEM faculty in Project Engage is three-fold. First, faculty provide oversight of the mentoring process through meeting monthly with their assigned mentors for updates, announcements, and discussion of any relevant issues. Second, STEM faculty act as professional role models for Project participants. Finally, the STEM faculty provide information on internships and other career guidance for both the mentors directly under them as well as the freshmen participants in the program. The STEM faculty play an important accountability role in the strategic mentoring process by ensuring that mentors fulfill their obligations. Having a three-tiered process provides an added measure of accountability to the process as well as an added source of support.

Peer Mentor Selection

Mentor selection is fundamental to the success of a mentoring program. The peer mentors chosen from upperclassmen, either juniors or seniors, are the first tier in the mentoring scaffold. Mentees need to perceive the mentor as supportive of their abilities and potential for success. Griffin et al. (2010) determined that regardless of racial similarity between the mentor and mentee, the mentor-mentee relationship was dependent on "the mentor's commitment to fostering the mentee's academic success" (p. 98). Accordingly, peer mentors in Project Engage are racially and ethnically diverse. Project Engage mentors undergo a rigorous selection process including an analysis of academic standing, leadership qualities, and faculty recommendations. Potential mentor candidates submit an application, curriculum vitae, and three professional letters of recommendation for review. Terrion and Leonard (2007) asserted that selecting a mentor who is not experienced and knowledgeable of university procedures compromises the mentor's ability to counsel the mentee in career-related matters as well as issues related to college-life. Additionally, mentees who share the same major as their mentor tend to be more satisfied with the mentee-mentor relationship. In order to qualify as a mentor for Project Engage, the applicants must be undergraduate upperclassmen STEM majors currently enrolled at the junior or senior level or graduate students in good academic standing. These qualities assure the mentor has the needed university experience, the needed experience in the STEM major, and is academically qualified to tutor a mentee if needed. Furthermore, providing a supportive environment for mentees can reduce anxiety and stress related to beginning a college career (Terrion & Leonard, 2007). Once the applicant pool is established, the Project Engage faculty team meets to evaluate all applications and select mentors representing all areas of STEM. Mentors must be dedicated to the fulfillment of their obligations to their mentees and to the program. The responsibility of overseeing mentee participation in all program events lies on the mentor. In addition to oversight of event participation, mentors are responsible for academic and/or social advising. In order to ensure the mentor-mentee interactions are supportive of the development of the mentees, Project Engage employs a graduate student who serves not only as the career counselor for the program, but is also charged with monitoring mentors' actions. Mentoring experiences can have long-term effects on success in STEM (Griffin et al., 2010); therefore, selection of high quality, responsible mentors and oversight of the mentor-mentee relationship are vital to the success of the mentoring program.

Mentor Training for Peer Mentors

After the selection of 10 to 14 mentors representing all areas of STEM, recruits undergo an initial training process. Mentors must complete the initial training meeting prior to meeting with their assigned mentees, with follow up meetings throughout the semester. Prior research has demonstrated the effectiveness of mentor training programs for improving communication skills, developing mentor-mentee expectations, and increasing mentor self-efficacy in regards to dealing with diversity issues (Johnson & Gandhi, 2015). Mentor training is very comprehensive involving multiple aspects of STEM education issues. The selected student mentors receive training related to their job expectations, university resources, cultural issues pertaining to minorities in STEM fields and specifically issues pertaining to rural students, basic listening skills, as well as peer counseling techniques prior to the beginning of the fall term each year. Additional mentor training sessions occur throughout the semester to update mentors and address any issues that arise. Often peer mentoring is the first type of leadership role these students incur, thus intensive training is required for successful mentor-mentee relationships to develop.

Peer Mentor Responsibilities

In addition to maintaining a 3.0 G.P.A. and being a currently enrolled junior, senior, or graduate student in good standing in a STEM discipline, each mentor is assigned five to seven mentees from the STEM freshmen seminar cohort. As mentors, the primary responsibility is the orientation and nurturing of STEM majors in their freshman year with the following responsibilities: introduce freshmen STEM students to STEM faculty, orient freshman Engage students to university policies and resources, share academic and personal success strategies with mentees, and conduct small group sessions in which students share their challenges/successes and receive peer support and guidance from their mentor. Mentors also tutor mentees on an as needed basis, but are primarily responsible for directing struggling students, with major academic or personal issues, to appropriate student support offices on campus. The student population primarily hails from rural areas and are often first-generation college students. They do not have the inherent knowledge of college life and available campus resources. Mentors play a vital role in their acclimation to college life as well as their STEM major. Project Engage mentors meet weekly with their assigned mentees, either individually or as a group, and document the topic of discussion for each mentee. After conducting their weekly meetings, mentors must upload a meeting checklist for each mentee in attendance to SharePoint, a secure online file sharing system. The documentation provides a measure of accountability and serves as an ongoing record of the mentee's progress.

Mentoring by the Graduate Career Counselor

The second level of scaffolding in the mentoring triad is provided by the program's career counselor. The career counselor provides multiple functions for the STEM program including mentoring, career counseling, and event organization. The career counselor serves as a role model and mentor for both the upperclassmen peer mentors as well as the STEM students as they progress through their college STEM programs. Applicants for the position must provide two letters of professional reference, a vita, and transcripts along with a completed application. The following describes the mentoring duties assigned to the career counselor:

- Provide immediate oversight of 10 to 12 undergraduate STEM mentors and all mentoring activities;
- Maintain documentation of the mentoring sessions;
- Conduct individual career/mentoring counseling sessions with program participants; and
- Assist the STEM faculty in mentor training sessions.

The STEM student participants, as well as the STEM mentors, look to the career counselor to provide guidance throughout the program while the STEM faculty depend on the counselor for direct mentor supervision. Having immediate contact with all levels of the Engage program participants, STEM students, mentors and faculty, the career counselor is an essential member of Project Engage.

STEM Faculty Mentoring

Completing the three tiered scaffolding of the mentoring model employed by Project Engage, is mentoring by STEM faculty. Many rural students are also minority, first-generation college students needing a supportive learning environment. Booker and Brevard (2017) examined the experiences of 58 African American students during their first year of college revealing that these students found mentoring beneficial to their transition to college. STEM faculty who take an interest in the welfare of their students provide the surrogate family relationship needed by rural, minority students to persist in a STEM major. Engage participants benefit academically and professionally through the development of relationships with the STEM faculty. STEM faculty are assigned mentors who have similar STEM interests in order to serve as role models and sources of information/resources for the upperclassmen mentors. Furthermore, STEM faculty teach the STEM 101 freshmen seminar classes allowing them direct contact with the freshman Engage participants. By meeting in class regularly with the STEM freshmen, deeper more long-lasting relationships are established and research collaborations form. STEM faculty complete the triad of mentors or MKOs building a comprehensive scaffolded approach to mentoring.

RESEARCH FINDINGS FROM A STRATEGIC MENTORING PROGRAM

Mentoring Survey Participants

To assess the effects of the mentoring program, online mentoring surveys were administered to STEM students enrolled in Project Engage. The survey data were available from the 2012 (n = 11), 2013 (n = 20), and 2016 (n = 10) cohorts. Of the 41 Project Engage STEM students who participated in the online mentoring surveys, about 73% were female; 66% were African Americans; and 54%, 15%, 12%, and 19% were respectively science, technology, engineering, and mathematics majors. The following table (Table 1) shows the breakdown of the demographic information of the respondents of the online mentoring survey.

Mentoring Survey Instrument

In addition to the four demographics questions about gender, race, ethnicity, and STEM majors, the mentoring survey instrument used for collecting the data also included 15 survey items. The 15 survey questions were positively worded to avoid confusion for respondents (Netemeyer, Bearden, & Sharma, 2003) and cognitive burdens for information processing (Dillman, 2006). The survey items adopted a 4 point Likert-like scale (i.e., strongly disagree,

Table 1. Demographics of mentoring survey participants

	2012 Cohort (n = 11)	2013 Cohort (n = 20)	2016 Cohort(n = 10)
Gender	Female: 8	Female: 16	Female: 6
	Male: 3	Male: 4	Male: 4
Ethnicity	Hispanic/Latino: 0	Hispanic/Latino: 2	Hispanic/Latino: 1
	Not Hispanic/Latino: 11	Not Hispanic/Latino: 18	Not Hispanic/Latino: 9
Race	African American: 7	African American: 12	African American: 7
	White: 4	White: 7	White: 2
		Mexican: 1	Hawaii & Pacific Islander: 1
Major	Science: 9	Science: 11	Science: 2
	Technology: 2	Technology: 2	Technology: 2
		Engineering: 3	Engineering: 2
		Mathematics: 4	Mathematics: 4

disagree, agree, and strongly agree). Among the 15 survey items, there are items intended to elicit responses regarding the participants' general impression of their experience in the Project Engage mentoring program, and there were items that focused specifically on assessing the mentoring program's impact in terms of academic performance, persistence as STEM majors, and intention to choose a STEM career after graduation.

Mentoring Survey Results

Overall Mentoring Program Effects

In analysis of the mentoring survey data, the students' responses to the 15 survey items were coded with "1 = strongly disagree", "2 = disagree", "3 = agree", and "4 = strongly agree". The means for the following three survey items were calculated to provide a general picture about the effects of the Project Engage mentoring program:

Item 1: The mentoring program of Project Engage was great.
Item 2: I enjoyed the mentoring program of Project Engage.
Item 3: I benefited from the mentoring program of Project Engage.

The means for the responses to three survey items are respectively 3.41, 3.64, and 3.56 indicating that, generally, the STEM students had positive experiences in the mentoring program and the program had positive effects on the students. The frequency distributions of the responses to the above three survey items by the 2012, 2013, and 2016 cohorts are shown in Figures 2-4.

The results presented in Figures 2, 3, and 4 show that, for the three survey items, 2013 Cohort's responses were less positive than those from 2012 Cohort and 2016 Cohort. Based on the well documented achievement gap between white and non-white students and the learning environment in rural areas, it was hypothesized that race makeup was related to the difference among the 2012, 2013, and 2014 Cohorts in their responses to the three survey items. To test this hypothesis, three Fisher's exact tests were conducted on three 2 x 4 contingency tables with 2 categories in race (i.e., white and non-white) and 4 categories representing the four possible responses of "strongly disagree", "disagree", "agree", and "strongly agree". The two students identified respectively as Mexican and Hawaiian & Pacific Islander in Table 1 were included in the "non-white" category in the contingency tables. Chi-square

Figure 2. Survey item 1 response frequency distribution

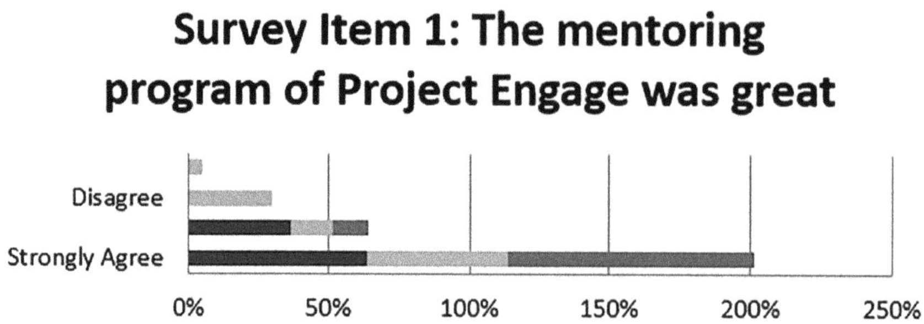

	Strongly Agree	Agree	Disagree	Strongly Disagree
2012 Cohort	64%	36%	0%	0%
2013 Cohort	50%	15%	30%	5%
2016 Cohort	88%	13%	0%	0%

Figure 3. Survey item 2 response frequency distribution

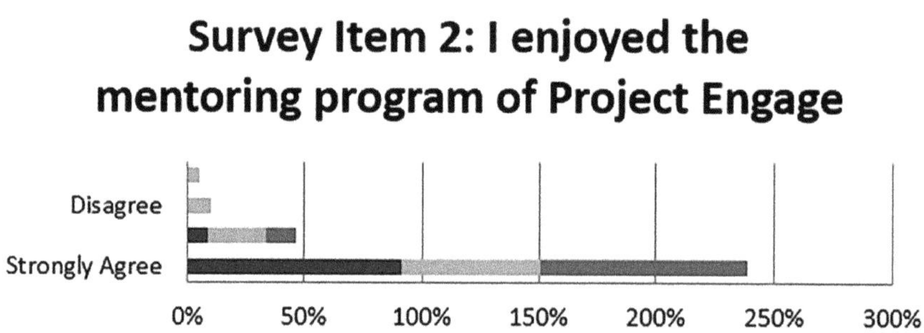

	Strongly Agree	Agree	Disagree	Strongly Disagree
2012 Cohort	91%	9%	0%	0%
2013 Cohort	60%	25%	10%	5%
2016 Cohort	88%	13%	0%	0%

tests of independence were not conducted because the expected value for some cells was less than 5.

For survey item 1, the Fisher's exact test results indicated a significant relationship between race and the four possible responses to the survey statement (p = .010). Among students who rated "strongly agree" to the statement of "the mentoring program of Project Engage was great," 16.7% were white students and 83.3% were non-white students. While 76.9% of

Figure 4. Survey item 3 response frequency distribution

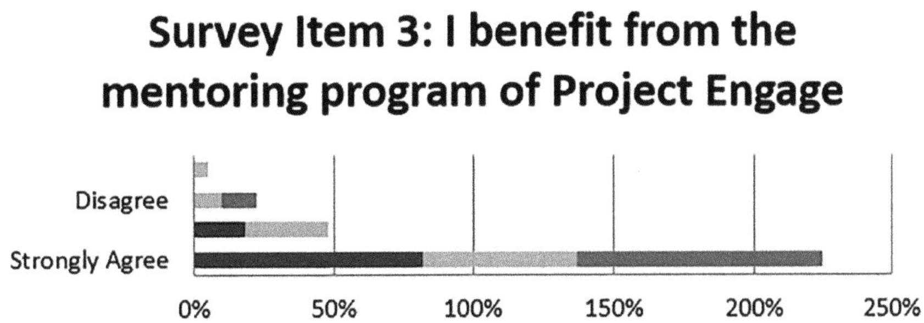

	Strongly Agree	Agree	Disagree	Strongly Disagree
2012 Cohort	82%	18%	0%	0%
2013 Cohort	55%	30%	10%	5%
2016 Cohort	88%	0%	13%	0%

non-white students rated "strongly agree," only 30.8% of white students rated "strongly agree."

The Fisher's exact test for survey item 2 did not yield statistically significant ($p = .340$) results indicating race was not related to students' rating of their enjoyment of the mentoring program. For survey item 3, the p value of Fisher's exact test was .049 indicating a significant relationship between race and student's rating of the benefit of the mentoring program. Among students who rated "strongly agree" to the statement of "I benefited from the mentoring program of Project Engage," 25.9% were white students and 74.1% were non-white students. While 76.9% of non-white students rated "strongly agree," only 53.8% of white students rated "strongly agree."

Impact on Academic Performance

Rural students are less academically prepared in STEM than their urban and suburban counterparts. It is, therefore, of great interest to examine effects of the Project Engage mentoring program on undergraduate STEM students' academic performance. Two relevant survey items for assessing such effects were:

- **Item A:** The mentoring program of Project Engagement helped you do better in school.

- **Item B:** The mentoring program of Project Engagement helped you do better in your STEM classes.

While item A focused on academic performance in general, item B looked particularly at academic performance in STEM. In analysis, the responses to the two survey items were coded with "1 = strongly disagree", "2 = disagree", "3 = agree", and "4 = strongly agree". The means for the responses to the two survey items are 3.18 and 3.13. The frequency distributions of the responses to the two survey items by the 2012, 2013, and 2016 cohorts are shown in Figures 5 and 6.

The results shown in Figures 5 and 6 indicate that the 2013 Cohort's responses to the two survey items were less positive than those of 2012 and 2016 Cohorts. To reveal if race was related to responses to survey items A and B, two Fisher's exact tests were conducted on two 2 x 4 contingency tables with 2 categories in race (i.e., white and non-white) and 4 categories representing the four possible responses of "strongly disagree", "disagree", "agree", and "strongly agree". The two students identified respectively as Mexican and Hawaiian & Pacific Islander in Table 1 were included into the "non-white" category in the contingency tables. Chi-square tests of independence were not adopted because the expected value for some cells was less than 5. The

Figure 5. Survey item A response frequency distribution

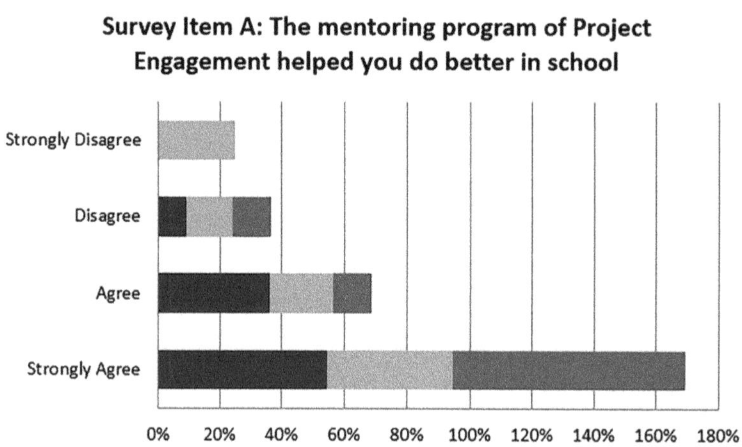

	Strongly Agree	Agree	Disagree	Strongly Disagree
2012 Cohort	55%	36%	9%	0%
2013 Cohort	40%	20%	15%	25%
2016 Cohort	75%	13%	13%	0%

Figure 6. Survey item B response frequency distribution

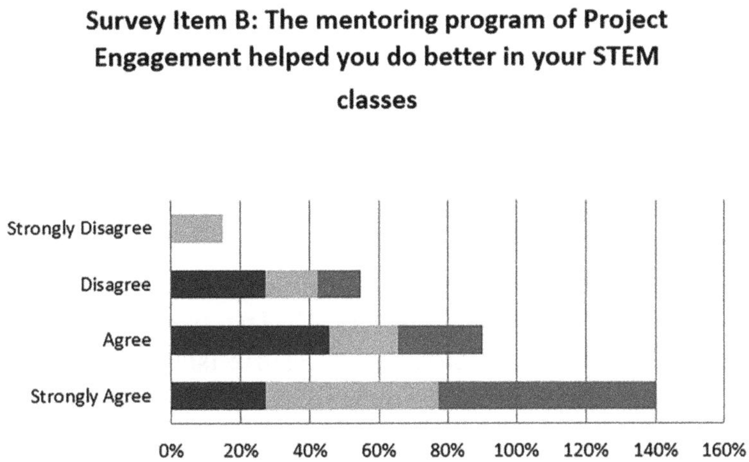

	Strongly Agree	Agree	Disagree	Strongly Disagree
2012 Cohort	27%	45%	27%	0%
2013 Cohort	50%	20%	15%	15%
2016 Cohort	63%	25%	13%	0%

results of the two Fisher's exact test were not statistically significant. But non-white students were more likely to rate "strongly agree" and "agree" to the two survey items than white students as indicated by the percentages of students in these two response categories: among those who rated "strongly agree" to survey item A, 25% were white students and 75% were non-white students, and the percentages of "agree" were 33.3% for white students and 66.7% for non-white students; among those who rated "strongly agree" to survey item B, 22.2% were white students and 77.8% were non-white students, and the percentages of "agree" were 36.4% for white students and 63.6% for non-white students.

Persistence as STEM Majors

One of the goals of Project Engage is to increase retention of undergraduate STEM majors. Therefore, it makes perfect sense to see if the mentoring program of Project Engage had a positive effect on retaining undergraduate majors. The responses to the survey item C "The Project Engage mentoring program influenced me to continue to be a STEM major in college" were analyzed. The responses to this survey item were coded with "1 = strongly disagree", "2 = disagree", "3 = agree", and "4 = strongly agree". The mean of the responses was 3.51 indicating a positive effect on retaining undergraduate STEM students. The frequency distributions of the responses to the survey item by the 2012, 2013, and 2016 cohorts are shown in Figure 7.

The frequency distributions in Figure 7 demonstrate that the three Cohorts showed similar pattern in their responses. To reveal if race was related to responses to survey item C, a Fisher's exact test was conducted on two 2 x 4 contingency tables with 2 categories in race (i.e., white and non-white) and 4 categories representing the four possible responses of "strongly disagree", "disagree", "agree", and "strongly agree". The two students identified respectively as Mexican and Hawaiian & Pacific Islander in Table 1 were included into the "non-white" category in the contingency tables. Chi-square tests of independence were not adopted because the expected value for some cells was less than 5. The result of the Fisher's exact test did not indicate a statistically significant result. But among those who rated "strongly agree" to survey item C, 26.9% were white students and 73.1% were non-white students, and the percentages for "agree" were 50% for both white and non-white students.

Figure 7. Survey item C response frequency distribution

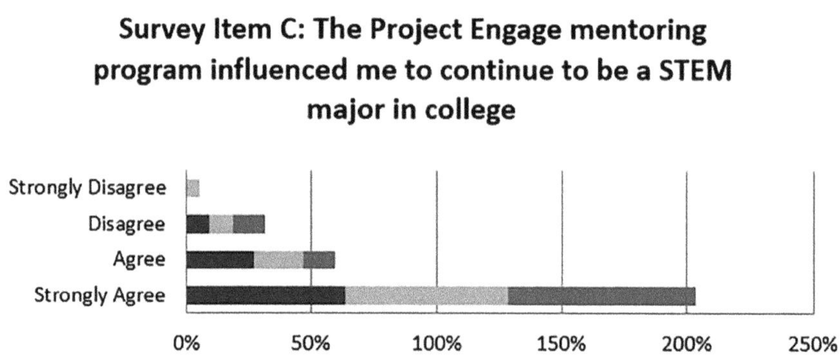

	Strongly Agree	Agree	Disagree	Strongly Disagree
2012 Cohort	64%	27%	9%	0%
2013 Cohort	65%	20%	10%	5%
2016 Cohort	75%	13%	13%	0%

Intention to Become a STEM Professional

To strengthen the STEM pipeline requires STEM majors to choose to work as STEM professionals after graduation. The responses to survey item D "the Project Engage mentoring program influenced me to continue to become a STEM professional after college" were analyzed to reveal if the Project Engage mentoring program had a positive effect on the undergraduate students' intention to become a STEM professional after graduation. The responses to this survey item were coded with "1 = strongly disagree", "2 = disagree", "3 = agree", and "4 = strongly agree". The mean of the responses was 3.46 indicating a positive effect on the undergraduate STEM students' intention to become a STEM professional after graduation. The frequency distributions of the responses to the survey item by the 2012, 2013, and 2016 cohorts are shown in Figure 8.

A Fisher's exact test was conducted on two 2 x 4 contingency tables with 2 categories in race (i.e., white and non-white) and 4 categories representing the four possible responses of "strongly disagree", "disagree", "agree", and "strongly agree" of survey item D. The two students identified respectively as Mexican and Hawaiian & Pacific Islander in Table 1 were included into the "non-white" category in the contingency tables. Chi-square tests of independence were not adopted because the expected value for some cells was less than 5. The result of the Fisher's exact test did not indicate a statistically significant result. But among the survey participants, the pattern that non-white students were likely to have more positive opinions about the effects

Figure 8. Survey item D response frequency distribution

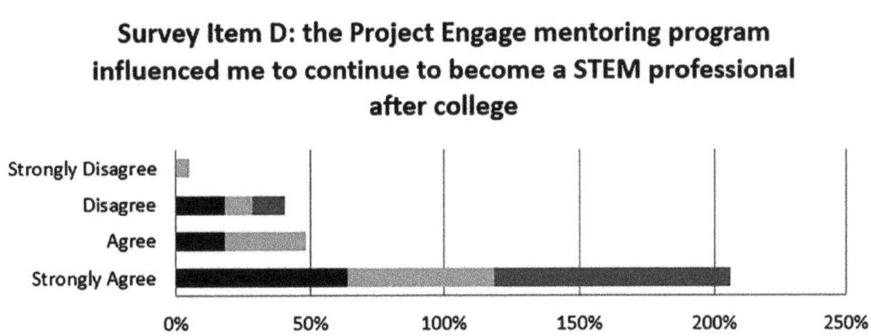

	Strongly Agree	Agree	Disagree	Strongly Disagree
■ 2012 Cohort	64%	18%	18%	0%
■ 2013 Cohort	55%	30%	10%	5%
■ 2016 Cohort	88%	0%	13%	0%

of the mentoring program on their intention to become STEM professionals was rather obvious with: 1) 76% of those rated "strongly agree" being non-white and only 24% being white; and 2) 73.1% of non-white students and 46.2% of white students in the "strongly agree" category.

SUMMARY AND FUTURE RESEARCH DIRECTIONS

The means of the 2012, 2013, 2016 cohorts' responses to the three survey items regarding the mentoring program's overall effects ranged from 3.41 to 3.64 indicating positive opinions of the students on the mentoring program. Among the survey respondents, there was an obvious trend that the non-white students were more likely to give higher ratings regarding the overall effects of the mentoring program than the white students. The Fishers' exact tests comparing the non-white and white students' responses indicated a significant relationship between race and survey responses regarding the "greatness" and "benefits" of the mentoring program: with higher ratings from non-white students. These significant results may be explained by the fact that rural students from minority communities are less likely to have experience with STEM professionals or have parents who can provide the knowledge needed to be successful in college and are, therefore, more responsive to mentoring programs.

Although the results indicated that non-white students were more likely to benefit from such mentoring programs as provided by Project Engage, there is no evidence-based information about what the specific benefits were and how much the benefits could be. Such information is essential to help future mentoring programs serve non-white or minority students better and research endeavors intended to furnish such information would result in great contributions to undergraduate STEM education in rural areas.

The means of the 2012, 2013, 2016 cohorts' responses to the survey items regarding the Project Engage mentoring program's impact on academic performance, persistence as STEM majors, and intention to become STEM professionals after graduation ranged from 3.13 to 3.51 indicating the positive effects of the mentoring program in these three aspects. Although the data indicated the non-white students were more positive about the mentoring program's effects on improving academic performance, persistence as STEM majors, and intention to become STEM professionals after graduation, the Fisher's exact tests did not indicate a significant relationship between race and responses to the three survey items. Therefore, the response difference noted in the non-white and white students in the 2012, 2013, and 2016 cohorts cannot be concluded among non-white and white students in general. Future research needs to test the relationship between race and the responses to the three survey items with more proportionate numbers of non-white students and white students. It is envisaged future research on similar mentoring programs will: 1) treat academic performance, persistence as STEM majors, and intention to become STEM professionals after graduation as dynamic variables that will evolve with time and 2) track students' trajectories of the three variables over time.

CONCLUSION

The three-tiered strategic mentoring process incorporated a social constructivist view and included the best aspects of both the community and hierarchical models of mentoring. Freshmen, predominantly minority STEM students, experienced a multi-tiered approach to mentoring consisting of mentoring by peers, graduate students, and faculty. Mentees were part of a larger cohort of STEM majors who benefited from the collaborative nature of belonging to the Project Engage community. The social interaction between mentor and mentee provided a foundation for freshmen mentees to develop STEM self-efficacy as well as the motivation and determination to persist in a STEM field. Using a scaffolding model including individual and group mentoring sessions along with social and academic interactions in the freshmen seminar course, STEM mentors (upperclassmen peers, graduate counselor, and faculty) served as the More Knowledgeable Other (MKO) elucidating the nuances of a STEM major. This scaffolded approach guided the undergraduate STEM major to their zone of proximal development (ZPD); enhanced motivation to pursue a STEM major, increased academic performance, and improved persistence rates.

The structured mentorship component of Project Engage is key to student success in STEM. As one of the four dimensions of the proposed STEM Education Framework, scaffolding via the mentoring process is key to supporting the development of a supportive learning environment and STEM self-efficacy. All three levels of mentors provide crucial support for young STEM majors. Results from the mentoring surveys revealed more positive responses from non-white students towards the mentoring program. Since Project Engage purposefully targets minority student retention, positive responses to the mentoring program were promising. Freshmen STEM majors need the experienced upperclassmen mentor to provide first-hand academic counseling and social advice; the graduate level career counselor for the ongoing career counseling as they progress through their undergraduate careers; and the faculty mentor as the STEM professional role model. The three tiered mentorship model also furnishes a method of accountability as each level is responsible to the next higher level from mentee, to mentor, to graduate mentor, to faculty mentor. As MKOs, mentors provide the scaffolded support to elevate the mentees to a higher level of understanding (their ZPD).

ACKNOWLEDGMENT

This research was supported by the U. S. Department of Education Minority Science and Engineering Program [grant number P120A120099] and the U. S. Department of Education Minority Science and Engineering Program [grant number P120A160122].

REFERENCES

Booker, K., & Brevard, J. E. (2017). Why mentoring matters: African-American students and the transition to college. *Mentor: An Academic Advising Journal*, 1. Retrieved from https://dus.psu.edu/mentor/2017/01/why-mentoring-matters-african-american-students-and-the-transition-to-college/

Bouwma-Gearhart, J. L., & Bess, J. L. (2012). The transformative potential of blogs for research in higher education. *The Journal of Higher Education*, *83*(2), 249–275. doi:10.1080/00221546.2012.11777241

Bouwma-Gearhart, J. L., Perry, K. H., & Presley, J. B. (2014). Improving postsecondary STEM education: Strategies for successful interdisciplinary collaborations and brokering engagement with education research and theory. *Journal of College Science Teaching*, *44*(1), 40–47. doi:10.2505/4/jcst14_044_01_40

Chester, A., Burton, L. J., Xenos, S., & Elgar, K. (2013). Peer mentoring: Supporting successful transition for first year undergraduate psychology students. *Australian Journal of Psychology*, *65*(1), 30–37. doi:10.1111/ajpy.12006

Dennehy, T. C., & Dasgupta, N. (2017). Female peer mentors early in college increase women's positive academic experiences and retention in engineering. *Proceedings of the National Academy of Sciences of the United States of America, 114*(23), 5964–5969. doi:10.1073/pnas.1613117114 PMID:28533360

Dillman, D. A. (2006). Why choice of survey mode makes a difference. *Public Health Reports (1974-), 121*(1), 11-13.

Galloway, C. M. (2001). Vygotsky's Constructionism. In M. Orey (Ed.), *Emerging perspectives on learning, teaching, and technology*. Academic Press. Retrieved from https://textbookequity.org/Textbooks/Orey_Emergin_Perspectives_Learning.pdf

Griffin, K. A., Pérez, D. II, Holmes, A. E., & Mayo, C. P. (2010). Investing in the future: The importance of faculty mentoring in the development of students of color in STEM. *New Directions for Institutional Research, 2010*(148), 95–103. doi:10.1002/ir.365

Holland, J. M., Major, D. A., & Orvis, K. A. (2012). Understanding how peer mentoring and capitalization link STEM students to their majors. *Career Development Quarterly, 60*(4), 343–354. doi:10.1002/j.2161-0045.2012.00026.x

Johnson, M. O., & Gandhi, M. (2015). A mentor training program improves mentoring competency for researchers working with early-career investigators from underrepresented backgrounds. *Advances in Health Sciences Education: Theory and Practice, 20*(3), 683–689. doi:10.100710459-014-9555-z PMID:25274417

Jonassen, D. H. (1999). Designing constructivist learning environments. In C. M. Reigeluth (Ed.), *Instructional-design theories and models: A new paradigm of instructional theory* (Vol. 2, pp. 215–239). Mahwah, NJ: Lawrence Erlbaum Associates.

Kendricks, K. D., Nedunuri, K. V., & Arment, A. R. (2013). Minority student perceptions of the impact of mentoring to enhance academic performance in STEM disciplines. *Journal of STEM Education: Innovations & Research, 14*(2), 38–46.

Kobulnicky, H. A., & Dale, D. A. (2016). A community mentoring model for STEM undergraduate research experiences. *Journal of College Science Teaching, 45*(6), 17. doi:10.2505/4/jcst16_045_06_17

Netemeyer, R. G., Bearden, W. O., & Sharma, S. (2003). *Scaling procedures: Issues and applications*. Thousand Oaks, CA: Sage Publications. doi:10.4135/9781412985772

Piaget, J. (1952). *The origins of intelligence in children*. New York: International Universities Press. doi:10.1037/11494-000

Piaget, J. (1959). *The language and thought of the child*. Chicago: Psychology Press.

Rogoff, B. (1990). *Apprenticeship in thinking: cognitive development in social context*. New York: Oxford University Press.

Terrion, J., & Leonard, D. (2007). A taxonomy of the characteristics of student peer mentors in higher education: Findings from a literature review. *Mentoring & Tutoring*, *15*(2), 149–164. doi:10.1080/13611260601086311

Vygotsky, L. S. (1978). *Mind in society: The development of higher psychological processes*. Cambridge, MA: Harvard University Press.

Wells, G. (1999). *Dialogic Inquiry: Towards a Sociocultural Practice and Theory of Education*. New York: Cambridge University Press. doi:10.1017/CBO9780511605895

Wertsch, J. V. (1985). *Vygotsky and the social formation of mind*. Cambridge, MA: Harvard University Press.

Wilson, Z. S., Holmes, L., deGravelles, K., Sylvain, M. R., Batiste, L., Johnson, M., ... Warner, I. M. (2012). Hierarchical mentoring: A transformative strategy for improving diversity and retention in undergraduate STEM disciplines. *Journal of Science Education and Technology*, *21*(1), 148–156. doi:10.100710956-011-9292-5

KEY TERMS AND DEFINITIONS

Career Counselor: Someone who provides guidance related to career opportunities, educational requirements, skills needed, and internship opportunities. Assists in helping students determine how to achieve their career goals.

Cognitive Development Theory: The development of human intelligence. Piaget proposed cognitive development in stages throughout one's lifespan. Vygotsky proposed that cognitive development was based on social interaction.

Community Mentoring: Pairs undergraduate STEM students with multiple mentors including peers, graduate students and multiple faculty from STEM areas forming a community or group of STEM scholars.

Graduate Level Career Counselor: Graduate student in a STEM discipline responsible for the oversight of the mentoring program.

Hierarchical Mentoring: Mentoring approach that pairs undergraduate STEM students with STEM peers, staff, and faculty.

More Knowledgeable Other (MKO): Someone or something having a better understanding or a higher ability level regarding a particular task, process, or concept. In the case of Project Engage, mentors, career counselors, and STEM faculty serve the role of MKOs for the undergraduate STEM majors.

Peer Mentor: Junior or senior level undergraduate or a graduate level STEM major who serves to guide STEM freshmen through their first year of college.

Social Constructivism: A theory of learning which expounds knowledge is constructed through interaction and negotiation among people under social cultural environment.

Zone of Proximal Development (ZPD): The level of cognitive development that the learner can obtain with the guidance of teachers or in collaboration with peers.

Chapter 4
A Three-Pillar Approach to Preparing Tomorrow's STEM Professionals:
Developing Knowledge, Abilities, and Ways of Working

ABSTRACT

STEM professionals need specific knowledge, abilities, and general ways of working to be successful. In this chapter, the authors identify a three-pillar approach for preparing future STEM workers including 1) knowledge of STEM careers and professionals, 2) abilities to work in the STEM fields, and 3) ways of working as STEM professionals. Additionally, the individual components and activities of Project Engage that address each pillar are detailed. Finally, this chapter also presents the results and implications discovered through survey research designed to ascertain the participants' opinions of the project activities and the impact of the activities on retention in STEM fields and on participants' desires to continue into STEM careers. The survey results uncover a trend of more positive responses of minority students towards project activities designed to prepare future STEM professionals. This trend calls for future, more in-depth examinations on the project activities and similar ones as a means to increase the number of underrepresented minorities in STEM professions.

DOI: 10.4018/978-1-5225-6341-9.ch004

INTRODUCTION

Unequal access to educational preparedness and opportunities in STEM continues to be perpetuated along racial, socioeconomic, gender, and geographic lines (U. S. Department of Education, Office of Innovation and Improvement, 2016). Rural students are not often exposed to STEM professionals and their work environments, hindering their professional aspirations. Project Engage intends to change that circumstance by promoting awareness of the ways STEM professionals work, increasing students' abilities to work in STEM fields, and exposing students to STEM careers. This three pillar approach adopted by Project Engage ultimately transitions students from college to the workforce by preparing them as tomorrow's STEM professionals. In this chapter, the three pillars of preparation are presented, the activities supporting the pillars described, and results from surveys of activities communicated.

PREPARING TOMORROW'S STEM PROFESSIONALS: A THREE PILLAR APPROACH

The three-pillar approach to preparing STEM professionals for working in tomorrow's STEM workforce incorporates: 1) knowledge of STEM careers and professionals; 2) abilities to work in the STEM fields; and 3) ways of working as STEM professionals (see Figure 1). Typically, in the rural landscape, young adults go directly into the workforce rather than seeking postsecondary education as illustrated by data from the nine county school systems within the Project Engage service area. On average 17% of high school graduates from this region attend college, with some county school systems having as low as 4% of their graduates attending college. Of those who attend college and earn an undergraduate STEM degree, only a small percentage return to their hometown areas due to lack of employment opportunities. According to Harmon and Wilborn (2016), "For the vast majority of students, pursuing academic subjects like mathematics as preparation for (STEM) occupations aligns with the need or desire to leave home for a more prosperous opportunity in an urban place" (p.26). Ultimately, this trend leaves few STEM role models for high school students to want to emulate. Together, the lack of STEM role models and exposure to STEM careers can partially explain rural students' lack of interest in STEM disciplines and motivation to pursue a STEM career. Without these key aspects early in a student's educational preparation, aspirations to become a STEM professional never develop. To improve STEM majors' retention rate and enhance motivation to pursue STEM careers after graduation, Project Engage adopted a three-pillar approach to prepare STEM majors to become tomorrow's STEM professionals (see Figure 1). Equipping STEM majors, especially those from rural areas who may lack significant exposure to STEM, with the knowledge of STEM careers, abilities to work as part of a team, as well as the techniques used by STEM professionals ensures these students are prepared to face the challenges of a STEM major and ultimately become a valued member of the STEM workforce.

Working in STEM fields involves applying STEM content knowledge to solve real world problems, to discover, and to innovate. STEM professionals have unique skill sets and ways of working to get jobs accomplished. Hossain and Robinson (2012) postulated, "Success in STEM requires both technical and non-technical skills and dispositions. Curiosity, the ability to think logically and creatively in problem-solving, communication skills and the ability to work in teams are all required to succeed in STEM careers" (p. 450). The three-pillar approach in Figure 1 was adopted to allow STEM students enrolled in Project Engage to gain insights about STEM careers and professionals and their ways of working, and to acquire skills to work

Figure 1. Three pillar approach for preparing tomorrow's STEM professionals

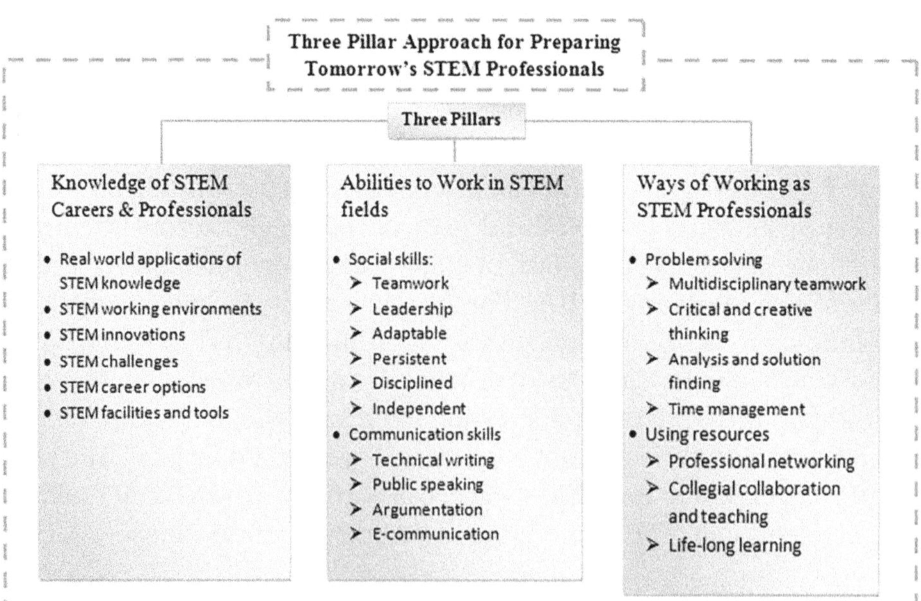

in STEM fields. This three-pillar approach was embedded in four types of Project Engage activities: STEM field trips, STEM Challenges and Invention Conventions, and invited speakers.

Through mastery experience and vicarious experience, Project Engage activities were intended to improve UWA STEM majors' confidence as future STEM professionals. The terms mastery experience (also known as enactive experience) and vicarious experience (also known as modeling) were identified by Bandura (1986, 1988) as two important sources of perceived self-efficacy. While mastery experiences are real and direct experiences that increase one's self-efficacy by allowing the individual to experience performance successes, vicarious experiences are indirect experiences that are able to increase one's self-efficacy beliefs by allowing the individual to observe similar others experiencing performance successes (Bandura, 1988). While the STEM field trips offered STEM majors mastery experience of real world STEM applications by exploring STEM facilities and workplaces and by interacting with STEM professionals in field trips, invited speakers activities allowed STEM majors to improve self-confidence through vicarious experience of interacting with STEM guest speakers or role models from STEM professions.

Hands-on experiences are a critical part of career exploration and skill building. Research has shown that hands-on learning activities not only provide students with more realistic and exciting experiences (Franklin & Peat, 2005), but also, lead to positive motivational outcomes (Thompson & Soyibo, 2002). Furthermore, hands-on experiences positively impact science achievement (Caglak, 2017). "Hands-on activities particularly present an avenue for facilitating and richening students' success and attainments in science" (p. 69). Guided by hands-on learning research, Project Engage organized STEM competitive events, STEM Challenges and Invention Conventions, for STEM majors and high school students from nearby school districts. Realistic and exciting experiences embedded in these activities were purposefully designed to help STEM students and high school students build social and communication skills required of STEM professionals and to practice ways of working as STEM professionals.

ACTIVITIES COMPRISING THE THREE PILLARS OF PREPARATION

Project Engage is a capacity building effort with the end goal of increasing the STEM workforce through comprehensive preparation of undergraduate students from rural areas, particularly those underrepresented in STEM. The approach has evolved into a three pillar methodology: increasing knowledge of STEM careers and professionals, cultivating abilities to work in STEM fields, and developing the ways of working as STEM professionals. To that end several components have been implemented in Project Engage: professional exploration field trips, invited STEM speakers, STEM competitive events (Invention Convention and STEM Challenge) which also includes an outreach to local high schools, and team building training. Together these activities provide rural students, both college and high school, with the needed knowledge, skills, and predispositions to pursue careers in STEM.

Professional Exploration Trips

The first step in preparing students to work as STEM professionals is to introduce them to a world of possibilities for STEM. Field trips to industries, government facilities, laboratories, nature centers, etc., allow students to explore STEM careers and learn first-hand from the professionals who work in these careers. Many students from rural areas of Alabama have no role models in the STEM professions and may not have traveled outside their rural counties. Factors such as isolated locales, lack of transportation, poor performing schools, lack of STEM industries in the areas, etc., contribute to the students' limited view of what a STEM professional is or does. Professional exploration field trips involve trips to STEM industry or other work sites where undergraduate students are allowed to interact with STEM professionals in their work environment. The onsite experience provides the students an opportunity to witness the STEM professional in action, thus, expanding their understanding of STEM professions and the opportunities that exist. Professional exploration field trips support the first pillar of the proposed three pillar approach to preparing tomorrow's stem professionals: "knowledge of stem careers and professionals". Field trips to STEM workplaces give students a glimpse of authentic STEM work environments, STEM professionals doing their jobs, and real world applications of STEM. Offering a variety of STEM facilities and professionals proffers maximum opportunity to capture students interest. Over the span of Project Engage, STEM professional exploration field trips have included the: 1) U. S. Space and Rocket Center, 2) Hudson-Alpha Institute for Biotechnology 3) Alabama Nature Center, 4) Marshall Space Flight Center/Redstone Arsenal, and 5) Mercedes-Benz U. S. International Incorporated. The variety of trips embodies all areas of STEM.

U.S. Space and Rocket Center

Year one field trip planning for Project Engage was a learning experience. The first field trip was to the U. S. Space and Rocket Center located in Huntsville, Alabama. The facility houses historical spaceflight memorabilia and displays including,1500 artifacts from America's space program as well as simulators where visitors can experience aspects of space flight. Visitors have access to the Rocket Park, Shuttle Park, International Space Station Exhibit, Museum Simulators, and other exhibits. Although docents led the Engage group through the historical space exploration displays, the trip lacked the personal interaction with STEM professionals in their work environment. Personal interaction is needed to foster student understanding of the role of STEM professionals in the workforce. In addition to visiting the historical displays and experimenting with the space flight simulators, Engage students completed the Parachute Lab, where they designed, built, and tested a parachute for effectiveness. The freshmen STEM students found the mathematical aspect of the parachute design very challenging. Mathematics is a challenging subject for most U. S. students, particularly those from rural areas where course offerings are minimal due to either a shortage of qualified math teachers and/or a scarcity of resources. The frustration of the students when trying to complete the Parachute Lab was indicative of the lack of mathematics preparation of our STEM student population. The trip was successful, as it fostered acquisition of content knowledge of the history of space exploration and engineering design skills needed for the Parachute Lab. However, it was missing a key component, exposure to STEM professionals in their genuine work environments.

Alabama Nature Center

For the second professional exploration trip, the Engage Project Team was careful to ensure that the program participants were able to interact with STEM professionals in order to gain an understanding of both the educational requirements and skills needed to achieve that career, and also to increase their awareness of STEM professions that they might not otherwise encounter. The Alabama Nature Center (ANC), the site of Project Engage's second career exploration field trip, is an outdoor educational facility located in Millbrook, Alabama on 350 acres of forestland with streams, ponds and wetlands (Alabama Wildlife Federation, 2017). The ANC staff worked with the Alabama Department of Environmental Management to create a customized career exploration program that introduced Engage participants to STEM careers in environmental science. Students were able to move through various stations where they seined a stream area to collect and identify invertebrates, learned to shock fish in order to do an assessment of the fish life in a pond, performed water quality chemical testing, hiked through a wetland observing the adaptations of the plants and animals in the wetland habitat, waded through a creek to learn about its inhabitants, and handled various wildlife. At each station an environmental professional introduced the career, the educational requirements, and the qualifications needed for a career in the field. Project Engage students were able to interact with the STEM professional performing actual tasks that the professional would do as part of their professional responsibilities. This field trip model was then replicated for future professional exploration trips to maximize student exposure to STEM professionals. The following list explains the ANC activities in more detail:

- Water Quality 101
    - **Overview:** Students determined water quality using both water quality test kits and by sampling the macroinvertebrate population.
    - **Purpose:** Explore how biologist, chemists and/or environmental engineers sample aquatic communities to determine different elements of ecological health; measure pond health at the ANC in one of two ways (chemical assessment or biological assessment); discuss relevance of factors being analyzed; and discuss why these factors are even more critical for agricultural fish production.
- Aquatic Life Roundup
    - **Overview:** Students determine water quality by sampling the macroinvertebrate population using nets and/or waders and the fish seine.
    - **Purpose 1:** Learn how environmentalist and/or biologists identify species richness, species diversity, adaptations of aquatic invertebrates, life cycles of amphibians and invertebrates,

metamorphosis, habitat preferences, predator-prey interactions, and test water quality via bio assessment.
    - **Purpose 2:** Explore how fish biologists maintain population control, population health, and agricultural industry.
- Wetlands Hike
    - **Overview:** Students explore the unique adaptations of wetland plants and animals.
    - **Purpose:** Learn wetland definitions, types, functions, plant life, wildlife, diversity, and conservation techniques employed by wetland scientists.
- Creek Culture
    - **Overview:** Students learn more about the habitat and life history of the animals that call the creek home.
    - **Purpose 1:** See who's on the ANC's Most Wanted list during a hike down Still Creek. Determine where the usual suspects are hiding out.
    - **Purpose 2:** Examine aquatic adaptations.
        - Determine how the wildlife adapt to live in changing freshwater environments.
        - See what anatomy (adaptation) allows them to cope with life in the water.
    - **Purpose 3:** See what fish are darting around Still Creek.
    - **Purpose 4:** Determine what creates a Creek?
        - From pools to riffles to runs, find out what makes a creek such an inviting place to live.
        - Learn more about the soil, rocks and minerals that create the creek itself.
- Alabama Animals
    - **Overview:** Students learn what animals inhabit the Alabama forests.
    - **Purpose:** Handle reptiles, amphibians and small mammals found in Alabama forests. Learn how STEM educators develop the curriculum for the ANC.

Marshall Space Flight Center/Redstone Arsenal

During years two and three the Engage Project Team collaborated with both the Marshall Space Flight Center/Redstone Arsenal and the Hudson-Alpha

Institute for Biotechnology to create a professional exploration field trip that exposed students to STEM careers in the aeronautical and biotech industries. The aim of this dual trip was to try and capture the attention of students with varied STEM interests. Survey results from the previous year indicated some dissatisfaction with the narrowly focused trips. Maintaining student satisfaction and interest in STEM is important for sustaining motivation to persevere and complete a STEM program. Thus, the Project Team sought to carefully plan trips so as to include all aspects of STEM.

The Marshall Space Flight Center, a National Aeronautics and Space Administration (NASA) facility, has designed rocket engines and launch vehicles, as well as other launch and payload systems for the advancement of U.S. space exploration. Today, Marshall is at the forefront of space exploration as it "is leading development of the Space Launch System, the most powerful rocket ever built to carry human explorers, cargo and science payloads farther into space than ever before, including a journey to Mars" (NASA, 2017, p.1). The educational staff at the Marshall Space Flight Center coordinated a robotics program of learning and competition, as well as a tour of the Redstone Arsenal. Additionally, the scientists introduced students to space flight projects, careers in the space flight control center as well as careers in materials science and engineering. Redstone Arsenal is a military installation that supports space exploration and U. S. missile defense. Moreover, Redstone employs scientists and engineers who are responsible for research, design, testing, and evaluation to further scientific and technological innovation. Students were able to tour the facility and interact with these scientists and engineers on a personal level. The STEM undergraduate students were allowed to hold metal replacement parts that had been 3-D printed allowing them to see science in action. Students were also allowed to explore the inside of model space shuttles and the command center where Redstone employees interact with individuals aboard the International Space Station. This trip provided first-hand knowledge of the possibilities for careers in the aeronautical field. The STEM professionals also provided information related to the acquisition and importance of internships. For some of the students with little knowledge and experience in STEM professions beyond medical professions, the availability of internships and their value to advancing a STEM career was eye-opening.

The second part of the trip focused less on specific careers and more on developing the knowledge and skills needed to work in STEM fields. In this portion of the experience science educators taught robotics and programming. During the first phase of the program, instructors provided a basic knowledge of programming. Teams were then charged with programming a robot to complete a specific task. Following the instructional phase, students were divided into teams and partook in a robotics competition that pitted the teams against each other. This STEM engagement activity served to further introduce and develop skills such as critical thinking, problem solving, and team building, needed to be successful in STEM careers. Overall, this trip provided a comprehensive look at the possibilities of careers in aeronautics while simultaneously engaging STEM students in an activity that taught needed knowledge, skills and ways of thinking by successful STEM professionals. The following list describes the activities of the Marshall Space Flight/Redstone Arsenal trip in more detail.

- Environmental Control and Life Support System (ECLSS) Facility
  - **Overview:** Marshall/Redstone contains the ECLLS facility and is responsible for the design, creation, and testing of the systems necessary to provide life support for the International Space Station including systems for oxygen generating, providing potable water, maintaining cabin pressure, and regulating temperature.
  - **Purpose:** Project Engage students explored the testing facility while discussing the individual systems with the project scientists
- Additive Manufacturing Facility
  - **Overview:** Additive manufacturing is a process by which a 3-D digital design image is fed into a 3-D printing machine which builds the design item by layering a material such as plastic thread or metals.
  - **Purpose:** Project Engage students were allowed to tour the facility, see the 3-D printer, handle the 3-D printed parts, and discuss the 3-D printing process with engineers at the facility.
- Payload Operations Integration Center
  - **Overview:** The Payload Operations Integration Center (POIC) serves the International Space Station as a communication center. All communications for the space station science operations experiments are managed by the POIC 24 hours a day.
  - **Purpose:** Project Engage students were not able to speak directly to those manning the communications systems, but instead were able to watch them in action as they communicated directly with the International Space Station. Watching the POIC communicate directly with the International Space Station was an inspiring experience.

- Robotics Training and Competition
  - **Overview:** The educational team at Marshall provided a robotics program including a general robotics lecture, instruction on programming, time for brainstorming and programming, as well as time for competition.
  - **Purpose:** Project Engage students participated in the four phases of the robotics program affording the opportunity for hands-on robotics activity. The robotics program challenged students to use their critical thinking and problem solving skills while working cooperatively as a team. With the team's success in accomplishing the robotics programming challenge, students were able to build their STEM self-efficacy. Increased STEM self-efficacy paves the road to retention and graduation.

HudsonAlpha Institute for Biotechnology

A tour of the HudsonAlpha Institute for Biotechnology was the third component of the year two and year three fall STEM professional exploration trip. The Engage Project Team purposefully planned this component to engage the students interested in biology and, in particular, genomics and biotechnology. The first component had solely focused on careers in mathematics, computers, and engineering geared toward the aeronautical industry. Many students had previously indicated that they preferred trips with a wider appeal. HudsonAlpha's mission is three-fold: to improve human health through genomic research, discovery and outreach to encourage future biotech scientists, and to educate the general public about genomic research discoveries. Since opening in 2008, HudsonAlpha has been responsible for genomic discoveries to benefit clinical medicine, promoting biotechnology industry growth, and educating the public and future members of the STEM workforce. One of the primary roles of HudsonAlpha is the advancement of genomic literacy. In one year alone, HudsonAlpha educates over 700,000 teachers and students about genomic science (HudsonAlpha Institute for Biotechnology, 2017).

Students from the rural areas of Alabama often recognize biology as a pathway to becoming a physician, but lack the vision to realize the multitude of careers possible with a biology degree. HudsonAlpha Institute of Biotechnology has an educational component where they introduce students to the world of biotechnology and genomics in a lecture-style setting. Engage students were informed about genomic science and its potential benefits to society. This introduction to the facility, its mission, and genomic science was then followed up with a tour of the various biotechnology laboratories. Students were fascinated with the laboratory equipment and were able to interact with the scientists as they worked on various projects. The one negative of this particular trip was the lack of a hands-on activity which would foster the skills needed, such as critical thinking and problem solving, for biotechnology careers.

- Genomics Education
    - **Overview:** A STEM education professional introduced the STEM freshmen to genomics as a scientific field of study and gave an overview of HudsonAlpha's purpose and scientific advancements in the field of genomics.
    - **Purpose 1:** Project Engage participants added an understanding of the study of genomics to their funds of knowledge.
    - **Purpose 2:** Project Engage freshmen learned of the many discoveries and advancements made by the researchers at HudsonAlpha, as well as the benefits of these discoveries to improving the human health condition.
- Genomic Laboratories Tour
    - **Overview:** Tour guides led the Project Engage students through a maze of laboratories providing ample time for interaction with the scientists as they worked.
    - **Purpose:** Exposure to advanced scientific research laboratories is lacking for STEM freshmen from underfunded educational settings. At HudsonAlpha, the STEM students were allowed to examine the complex laboratory equipment as they questioned the scientists about how they used it in their work.

Mercedes-Benz U.S. International Inc. (MBUSI)

Mercedes-Benz U.S. International Inc. (MBUSI) was the state's first auto manufacturer to establish in Alabama and employs approximately 4000 workers directly with another 18,000 individuals employed indirectly at factories which provide auto parts and supplies. The auto manufacturer produces three vehicle models: e M-Class SUV, the R-Class Grand Sports Tourer, and the GL-Class luxury SUV (Lewis, 2015). The plant works like a "well- oiled machine," from Body, Paint, and Assembly shops to Administration and

Shipping. As with most STEM professions/companies, teamwork is key to success. The organizational structure of the plant, having all of the aspects of the manufacturing process in one location "helps to emphasize the importance of teamwork and to foster Team Member communication..." (Mercedes-Benz U.S. International, Inc., 2017, para 1).

- MBUSI Facilities Tour
  - **Overview:** Students toured the MBUSI facility witnessing STEM in action as they observed automobiles being assembled. Students were able to see a STEM workforce interacting with the robotic machines involved in auto assembly.
  - **Purpose:** Project Engage students, being primarily from rural areas of Alabama, are not exposed to the complexities of STEM work environments. The tour of MBUSI gave students the opportunity to observe a real world application of robotics.

Again, as with the previously described professional exploration trips, students engaged with the STEM professionals in their workplace environments. Students were able to witness real-world applications of the science, technology, engineering and mathematics principles that they were learning in the classroom setting. Exposure to real world applications of STEM in addition to exposure to STEM careers and STEM professionals is essential for "experiencing STEM" in order to build self-efficacy and motivation. A report by Change the Equation and the AMGEN Foundation (2016) statistics revealed that a majority of teens claimed having an adult STEM role model in their field of interest would be beneficial (86%) yet only 32% actually had that connection. Furthermore, the 2016 Change the Equation and AMGEN Foundation report denoted the following:

Low-income students have the fewest pathways to science careers. They are less likely than high-income students to:

- *Know someone who works in biology (19% vs. 25%),*
- *Be interested in biology (69% vs. 78%),*
- *Volunteer in a career area that interests them (18% vs. 24%), and*
- *Get tested for career aptitude (23% vs. 29%) (2016, p.4).*

Project Engage worked with students from some of the poorest and most rural counties in Alabama. The statistics provided above illustrate the unequal access to STEM professionals and STEM career resources. Inequities such as these only serve to perpetuate the cycle of poverty and limit the pool of applicants for the U. S. STEM workforce. U. S. students are not being prepared or motivated to enter the STEM workforce (Hossain & Robinson, 2012; National Academy of Sciences, National Academy of Engineering, and Institute of Medicine, 2007). The U. S. STEM workforce is increasingly being supplied by foreign-born workers (Hossain & Robinson, 2012). To counter this trend, there needs to be a focus on preparing U. S. students to fill the STEM pipeline from college to career. Through career exploration field trips to STEM related businesses, exposure to STEM professionals, and opportunities to experience real world STEM applications, Project Engage is providing the framework for developing a stronger STEM workforce.

Competitive STEM Events

Professional exploration field trips exposed Project Engage students to many possible careers in STEM; however, it is the hands-on STEM competitive events that support pillars two and three for "Preparing STEM Professionals" by developing abilities and ways of working as STEM professionals. Students from rural, predominantly minority communities often leave high school and enter college lacking the motivation, self-esteem, and skills to be successful in a STEM major. Project Engage, like STEM 2026 has a vision for creating an atmosphere of excitement for lifelong learning though formal and informal STEM learning opportunities. In STEM 2026, a proposal for STEM education for preschool through twelfth grade (P-12) and beyond, the goal envisions "The language and concepts of STEM are accessible to all, and there is a shared understanding of where STEM fits into all people's lives, regardless of race or ethnicity, disability, language spoken, gender, neighborhood, or geographic location" (U. S. Department of Education, Office of Innovation and Improvement, 2016, p. 25). Both college freshmen and high school students quickly engage when competition exists. Project Engage capitalizes on this excitement over competition to teach STEM techniques, inspire, and build self-esteem. Through team-based competition, students develop essential social skills such as teamwork, leadership, adaptability, persistence, discipline, and independence as well as communication skills like technical writing, public speaking, e-communication, and argumentation. Competition also develops problem-solving skills and the ability to use resources efficiently. Two types of events, the Invention Convention and the STEM Challenge are held annually and involve both local high school students (to spark interest early) as well as the Engage freshmen participants (to motivate and promote persistence) as competitors. Initially, the high school and college teams competed against each other, but the Project Team quickly determined that both groups were dissatisfied with this practice. After year one, the college freshmen took on a leadership role in the high school student competitions and participated in competitive events only against other college students.

Invention Convention: Robotics Year 1 Lessons Learned

The focus of Invention Convention is on design and development. This STEM competition has taken on many different themes including robotics, egg-drop carriers, boomerangs, space rovers, and so on. The first competition to be hosted by Project Engage was a robotics competition. Robotics integrates all areas of STEM giving students the opportunity to develop the scientific, technological, engineering, and mathematical skills to prepare for future STEM careers. Moreover, the cooperative learning nature of working with robotics has been shown to increase critical thinking skills and interest in STEM fields (Mosley, Ardito, & Scollins, 2016), adding to the support of using robotics for preparing students to work in STEM fields. Robotics competitions require students to work in teams, develop leadership skills, and communicate effectively providing the ideal STEM learning experience.

During year one of the Project, Engage, freshmen participated in a robotics competition while high school students were challenged to invent a product by recycling something old. For the college students, the Project Team sought to capitalize on the current interest in robotics and the positive outcomes that can be achieved through robotics. The Engage freshmen were taught the concepts and skills needed for building and programming robots both within their freshmen seminar course, STEM 101, as well as during outside sessions working one-on-one with the Project Engage Computer Information Systems (CIS) professor. The CIS professor scheduled visits to the STEM 101 classes to introduce the freshmen to both the Lego Robotics kits used for building the robots and to logical thinking needed for programming the robots to perform a specific task. Robotics competition teams consisted of Engage freshmen and were led by Engage upperclassmen mentors. Additional sessions outside of class were needed to introduce the mentors to the same robotics building and programming skills as well as to allow scheduled work time. On the day of the competition, robots were evaluated on their complexity as well as their ability to complete their programmed task. Each team had to first present their robot and describe the task it was programmed to complete and follow up with a demonstration of the robot completing the intended task. Teams were scored on the complexity of the programming along with the results of the task performance. Freshmen not only learned the skills of building a robot with working parts and programming, but also learned problem-solving and critical thinking as they tested, revised, and re-tested the robots to ensure they were working optimally. Another important skill learned was team-building. STEM professionals typically do not work in isolation, but rather as part of a larger project team. The Project Engage leadership identified that teamwork skills in combination with leadership skills were essential to supporting the development of a working STEM professional. As a result, an outside consultant was brought in and taught team-building skills during the 101-STEM course, and students were then able to put into practice the newly acquired knowledge and skills as a member of the robotics competition team.

Invention Convention: Robotics Competition With the Experts

After the first robotics competition developed, taught, and hosted by the Engage Project faculty, it was determined that the competition would be better hosted offsite by a team of robotics experts. The time and difficulty scheduling class visits to the STEM 101 classes combined with the difficulty in adding robotics building and testing into an already busy student schedule resulted in some of the teams not being prepared for the competition. The goal of the competition was to increase interest in STEM fields and motivation to pursue a STEM career. The frustration expressed by some of the students, led the team to conclude that changes were needed. Thus, the Project Team worked with the education staff at Marshall Space Flight Center to develop a robotics competitive event for the Engage freshmen hosted on site at Marshall. In so doing, the freshmen were taught robotics building and programming by experts in the field of aeronautical engineering. Having dedicated time for the robotics training and competition resulted in all teams having equal access to the professionals and equal time for building and testing their robots. Student confidence in their abilities developed, as they experienced success without the frustration of trying to fit the robotics training and building into their already busy freshmen schedules.

The robotics program developed by the STEM education specialists at Marshall consisted of four phases: introduction to robotics, introduction to programming, design/building/programming, and competition. Throughout the introduction to robotics and programming stages, the teams received a firm foundation in the knowledge and skills needed to accomplish the robotics challenge. During the next phase (design/building/programming), members of the Engage student teams worked collaboratively on the robotics challenge, simulating the actual work experience of a STEM professional. The students were challenged by the complexity and unfamiliarity of the building and programming of robots, but met the challenge confidently due to the structure of the event as well as the knowledge of the experts leading the training. Finally, the teams were put into a head-to-head competition sparking enthusiasm and building STEM self-efficacy as teams successfully completed the robotics challenge. Due to the success of the robotics competition developed and hosted by the experts at Marshall, the Engage Project Team scheduled the subsequent robotics events at the Marshall Space Flight Center.

STEM Challenge: Forensics Year 1

The Project Engage Team sought to take advantage of the STEM competitive events as motivational activities and to maximize opportunities for students to increase the skills needed for working as STEM professionals. The spring competitive event during year one was a STEM Forensics Challenge created and hosted by the Engage Project Team. The Team wanted to capitalize on popular television programming which is saturated with crime scene investigation shows. Taking advantage of the popularity of crime scene investigation media is not an uncommon practice in outreach programs. Menshew used popular media themes of Crime Scene Investigator (CSI) and zombies to help students make connections with science principles (2015). Menshew's program has been successful as demonstrated by a comparison of California Standardized Testing scores between program participants and a control group. Students in the high school program outperformed their non-program peers. According to Menshew, "Numerous graduates have reported real-world academic value to their participation in the program. Scholarship and college/university applications are enhanced by program participation" (2015, p. 69). Youth are intrigued by popular television programming, thus, taking advantage of that interest engages students in STEM learning and promotes STEM self-efficacy.

Furthermore, rural students have a working knowledge of the importance of forensics and also see the "relevance" of forensics to their everyday lives and society in general. Stuckey, Hofstein, Mamlok-Naamanb, and Eilksa

(2013) described "relevance" as a complex concept but one important for student learning. Science learning is relevant for students when it addresses student interests and has positive outcomes for students, such as intellectual development, fostering student's understanding of future career opportunities, furthering student realization of their potential to contribute to society. Context-based curricula, such as forensics, promote learning in authentic situations which leads to students seeing the "relevance" of the science and having the ability to apply the learning to their future endeavors.

Crime scene investigations encompass all areas of STEM (science, technology, engineering and mathematics), thus, a STEM competitive event was developed for Project Engage with a forensics theme. During year one of the Forensics STEM Challenge, both college freshmen and high school students competed on teams against each other which resulted in dissatisfaction on both sides. The high school students felt they were at an academic disadvantage, while the college freshmen felt the Engage leadership did not have high expectations of their STEM knowledge and skills. Steps were taken during year two of the Engage grant to ensure that the high school students and the college freshmen were not pitted against each other in the competitive events. However, for year one the high school teams and the college teams were in direct competition.

The Forensics STEM challenge event started with a STEM warm-up which involved a series of science questions, a technology introduction to making observations, an engineering activity where students were required to create a scaled drawing, and a series of math questions related to forensics. The activities were as follows:

- **Science:** The science warm-up consisted of a series of science questions related to crime scene investigation. Questions were posed related to blood samples, blood typing, DNA knowledge, and hair samples.
- **Technology:** To incorporate technology into the STEM warm-up, the Project Team used an online web resource which required students to make crime scene observations. Several web resources were available to inspire students to pursue forensics, assist with developing investigation knowledge and skills, and provide an overall examination of the forensics profession.
- **Engineering:** Crime scene investigators are often required to construct models to aid in solving crimes. To develop skills in design and construction, participants were given an engineering design challenge. The engineering challenge consisted of one team member drawing an

object to scale and the second team member reconstructing the object from the scaled drawing.

- **Mathematics:** The math challenge, keeping with the forensics theme, consisted of a series of mathematical questions which required students to extrapolate the height of a suspect from their footprint/shoe size.

The warm-up activities were scored and used to ascertain the order that teams would start the forensics investigations. The highest scoring teams in the warm-up would begin first, thus, having more time for the investigation.

After the STEM warm-up, the teams met the "judge" and were introduced to the "crime," a hit and run accident that left a campus student dead. The students (college freshmen and high school) were instructed that they would be collecting evidence at various locations, including two outside locations and one laboratory, and that every argument or statement made needed to be supported by the evidence collected. The students then had to locate the various crime scenes across campus using coordinates and a compass. Using iPads, the students were required to take photos of the evidence to use in their presentations to the judge. Measurements, such as footprint length, were used to deduce the height of the suspect. All aspects of STEM were incorporated into either the crime scene itself or the tools used by the STEM student CSIs. When students arrived in the auditorium to present their cases, a judge, bailiffs, and suspects were all present, giving the challenge an authentic look and feel. "Investigators" were required to present evidence that supported the suspect or suspects identified as the perpetrator(s), identify a motive, and determine if the crime was accidental or premeditated. This activity required students to use scientific observation and data collection skills, logical reasoning to ascertain the identity of the culprit from a series of evidence and suspect interviews, as well as written and oral communication skills – all skills needed to be successful as STEM professionals. The Forensics STEM Challenge took advantage of a popular topic to teach various STEM techniques while simultaneously igniting student interest in and motivation to pursue a STEM major and subsequent career.

The Engage Project Team took results from the surveys administered the previous year to improve the Invention Convention and STEM Challenge activities for year two. The high school and freshmen college student teams' participation in the same competitive events was the most frequently noted item of dissatisfaction by both groups. In light of that dissatisfaction, the Engage Project Team decided to hold one event where the college freshmen teams competed and a separate event for the high school teams for the Invention

Conventions. Additionally, the STEM Challenges evolved into competitions for the high school students with opportunities for the college students to increase confidence in their STEM abilities as they led the high school teams and served in a mentoring capacity or other event leadership roles.

Invited STEM Speakers

Guest speakers who are experts in STEM fields support the first pillar of the "Preparing Tomorrow's STEM Professionals" model and are particularly important for reaching students from rural areas. Rural areas are often subjected to extreme isolation from the outside world. As a result, rural students lack access to a network of professionals in all STEM fields, especially a lack of minority STEM professional role models. Actively engaging students in a motivational program, such as Project Engage, that introduces STEM professionals helps students from these isolated rural areas succeed in STEM disciplines and become role models for their communities. Project Engage has sought professionals who represent a wide range of STEM fields including a NASA Project Manager, a former U.S. Surgeon General, president and founder of the Minority Aviation Education Association (MAEA) and owner of Interactive Science Programs, a physicist and world champion boomerang thrower, a rocket scientist/STEM motivational speaker, and a forensics expert from the Alabama Department of Forensics. The Project Engage Team carefully selected a variety of STEM speakers so as to give college and high school project participants a broad picture of STEM fields and the people who work in these professions. Invited STEM speakers help prepare future STEM professionals by informing students of the assortment of STEM career options. Getting to hear first-hand the rewards of working in a STEM field also serves to motivate students to continue on their STEM career paths.

Speakers from governmental agencies are appealing due to both their expertise and no cost. Mr. Todd May, current Director of NASA's Marshall Space Flight Center and former Program Manager for NASA's Space Launch System provided a look into the career of a scientist with NASA (NASA, 2016). He shared experiences working on numerous projects for the Science and Mission Systems Office at Marshall Space Flight Center including the development of the Space Launch System, the integration, launch and commissioning of the International Space Station's Quest airlock, and the Discovery and New Frontiers Programs (NASA, 2016). For many members of Project Engage, this was their first opportunity to interact with a STEM professional beyond the confines of the institution. The experiences shared by Mr. May, a native of Alabama, allowed the Project Engage participants to see their own potential. If an Alabama native could achieve such success in the STEM field of aeronautics, they too could be successful in the STEM field of their choice.

Dr. Regina Benjamin, the 18th U. S. Surgeon General, inspired both Engage participants as well as the community. Dr. Benjamin's stellar academic background and professional career achievements gave students a glimpse into the potential success that someone from a small rural area in Alabama could achieve. Dr. Benjamin earned a B.S. in chemistry from Xavier University of Louisiana, an M.D. degree from the University of Alabama at Birmingham and an M.B.A. from Tulane University. She was the first doctor under the age of 40 and the first African American female to serve on the American Medical Association Board of Trustees (U. S. Department of Health and Human Services, n.d.). Perhaps the most telling aspect of Dr. Benjamin's life story was the fact that she came from a rural area of Alabama (like Engage students), earned her STEM degrees, had a very successful career and then returned to her home to provide medical services for those in need. Engage students had the opportunity to see the potential for their lives through the life and successes of Dr. Benjamin.

Dr. Darryl Baynes, co-founder of Minority Aviation Education Association (MAEA) that was originally intended to encourage minority youth to pursue careers in the field of aviation, but was later extended to add a focus on STEM careers in general (Baynes, 2003), was invited to showcase his hands-on science program for both the surrounding rural schools and the undergraduate Engage participants. Dr. Baynes used an active learning methodology for teaching and learning of science upholding the philosophy that students learn science by "doing" science. He was a dynamic science teacher and entertainer, again demonstrating to the students from isolated rural areas a successful African American scientist role model.

Mr. Kantis Simmons, rocket scientist, author, and STEM speaker (Simmons, n. d.) provided a motivational message of encouragement and strategies for success in STEM fields. Again, Mr. Simmons was another example of a successful African American STEM professional which students from the rural Black Belt do not often have opportunities to observe. Mr. Simmons organized and led an Invention Convention competitive event for high school students. Additionally, he shared his strategies for being successful in STEM with the high school teachers and Engage college participants.

Mr. Justin Sanders, Chemistry Section Chief for Alabama's Department of Forensic Sciences (ADFS) Drug Chemistry Division provided expert testimony of the rewards of a career in forensics. In this division, he oversees the analysis of suspicious materials in order to identify controlled substances, non-controlled substances, precursor chemicals, and other chemicals used in the manufacture of illegal drugs. Mr. Sanders has contributed his expertise to the Forensics STEM Challenge creating an authentic learning experience.

Ware and Stein (2013) postulated "Mentors and role models can play a significant role in high school students' motivation to pursue specific careers later in life…" (p. 209). Invited STEM speakers (i.e. role models) provide Project Engage students and area high school students with the opportunity to interact with and learn from STEM professionals typically not available to students from rural areas. The successful careers of the speakers inspire and motivate rural students to continue their path towards a STEM career. This opportunity for interaction with successful STEM professionals has contributed to the development of the students' perceptions of the potential for a career in STEM.

Team Building Training

The complex nature of STEM related jobs demands that STEM professionals be able to work as part of a collaborative team. Working as a team member fosters communication, problem-solving, negotiation, cooperation, conflict resolution, decision-making, etc. (Allan, 2017). Freshmen do not always come to college knowing how to successfully work as a part of team, and yet, most colleges do not teach teamwork as an academic offering. Explicit training is needed to develop the necessary understanding and skills to be a contributing team member. Learning to work as a team member is critical to preparing to be a STEM professional. During the first two years of Project Engage, an outside consultant with expertise in team building was brought to campus. The consultant led each freshman STEM 101 class in an interactive exchange in what it takes to make an effective team. A key element in the training was creating team contracts outlining the responsibilities of each team member. Courtright, McCormick, Mistry, & Wang, (2017) "… draw on macro organizational control theory to propose that team charter quality serves as a team-level 'behavior' control mechanism that builds task cohesion through a structured exercise" (p.1462). Quality charters or contracts outlining each member's responsibilities, positively impact the success of the team in accomplishing its task, specifically teams with members who are less conscientious (Courtright et al., 2017). The team building expert was vitally important to the Engage program as most freshmen participants from the local rural area were first generation college students lacking many of the tools to be successful in STEM major fields including teamwork skills. Instruction was provided on both team effectiveness and the importance of developing team contracts when working on group projects. In addition, the expert stressed communication and the need for rules/guidelines when working in groups in order to create an effective team that can accomplish a goal successfully. The Project Engage student mentors and the career counselor were trained to expedite team work so that in future endeavors they could facilitate team projects.

Later in the term, the Project Engage Team, mentors, and student participants were able to make use of this team effectiveness training and team contracts to facilitate the work on a Lego Robotics competition which was part of an Invention Convention. Freshmen STEM students, guided by the rules of the team contracts, worked in teams to successfully design, build, and demonstrate a robot for competition. The teamwork training was a valuable asset in this robotics competition and will continue to benefit these students as they embark on STEM careers. Teamwork achieved through group projects increases the ability of students to work with diverse individuals collaboratively to achieve a goal. Participants in teamwork also benefit from developing their own teamwork skills as well as the understanding of the importance of being a part of a team (DiLisi, Eppell, & Upton, 2006).

The class-based teamwork instruction was beneficial to the members of Project Engage; however, the Project Team wanted to expand the teamwork training and incorporate a leadership component. The Alabama 4-H Center offered several activities including team building, the clover hop, the climbing wall, the giant swing, and a challenge course. Each spring Project Engage took a field trip to the facility where Engage students underwent intensive team building and leadership training as well as reflection on the activities. Descriptions of the activities at the Alabama 4 H Center are as follows:

- **Team Building:** Numerous team building activities are offered including "pass the can," "alligator pass," "blind polygon," "birthday line-up," "turnstile," and "raccoon circle." The main objectives of the activities are to develop cooperation, decision-making skills, problem-solving skills, as well as to foster trust between the team members, and develop self-confidence in the individual. Two of the team building activities performed by Project Engage are the "pass the can" and "birthday line-up."

  - **Pass the Can:** The team must pass a metal can around a circle without letting it fall to the ground. This activity requires participants to work collaboratively. It builds communication and problem-solving skills as team members work to ensure the can successfully travels around the circle without dropping to the ground.
  - **Birthday Line-Up:** The goal of "birthday line-up" is to have everyone line up while balancing on a log by birthday from youngest to oldest. The catch is that they cannot speak, they are blind-folded and if someone falls off the log, the group has to start over. This activity requires an extreme amount of teamwork as well as problem-solving.
- **Clover Hop:** The clover hop consists of a series of suspended platforms (clovers) attached to ropes. Participants must make their way across the series of clovers.
- **Climbing Wall:** The climbing wall is a 25 foot wall that teams must climb. The climbing wall challenges individuals and builds team spirit.
- **Giant Swing:** Participants are harnessed into a swing 35 feet above ground. The purpose of the activity is to build trust in teammates.
- **Challenge Course:** The challenge course is a series of difficult ground-based activities such as low rope challenges and a wall to climb over. Teams must work together to be successful in completing the challenges.

Most design and development projects are not accomplished individually, but require expertise from several team members. The Pillar of "Abilities to Work in STEM Fields" in Figure 1 outlines social skills (e.g., teamwork and leadership) and communication skills as the two primary categories of the abilities needed by STEM professionals. Incorporating specific teamwork and leadership training as components of Project Engage was important scaffolds to the development of students' abilities to work as STEM professionals.

RESEARCH FINDINGS FROM ACTIVITIES OF PREPARING TOMORROW'S STEM PROFESSIONALS

Field trips, STEM Challenges, and Invention Convention were the activities in Project Engage to prepare the undergraduate STEM students to become tomorrow's STEM professionals as delineated in Figure 1. Survey data were collected after these activities and were analyzed to reveal the effects of these activities.

Field Trip Surveys

Field trip survey data were available from 2012 (n =60), 2013 (n = 14), 2014 (n =26), 2015 (n =22), and 2016 (n = 28) cohorts. The demographic information for the survey participants is shown in Table 1. The 2014 and 2015 field trips were team-building field trips, and specific information and data analysis results of the two field trips are reported in the "Team Building Alabama 4-H Center Survey" section in this chapter. The instrument used to collect 2012, 2013, and 2016 field trip data consisted of four demographics questions about gender, race, ethnicity, and STEM majors and 8 survey items regarding the effects of the field trips on the undergraduate students' knowledge, skills, interest in STEM and STEM careers. The 8 survey questions were positively worded to avoid confusion for respondents (Netemeyer, Bearden, & Sharma, 2003) and cognitive burdens for information processing (Dillman, 2006). The survey items adopted a 4 point Likert-like scale (i.e., Strongly disagree, disagree, agree, and strongly agree).

2012 Field Trip to the U.S. Space and Rocket Center

The analysis of the 2012 field trip data focused on effects of the trip on undergraduate STEM students' competency to be successful in STEM, knowledge of STEM fields, skills of working on STEM projects, persistence as STEM majors, and intention to work as STEM professionals after graduation. In the survey data analysis, the responses were coded with "1 = strongly disagree", "2 = disagree", "3 = agree", and "4 = strongly agree".

Competency to Be Successful in STEM

The mean of the responses to this survey item was 2.93 not indicating a very positive rating regarding the field trip's effects on improving students'

Table 1. Demographics of field trip survey participants

	2012 (n = 60)	**2013 (n = 14)**	**2014 (n = 26)**
Gender	Female: 39	Female: 11	Female: 8
	Male: 21	Male: 3	Male: 18
Ethnicity	Hispanic/Latino: 2	Hispanic/Latino: 0	Hispanic/Latino: 2
	Not Hispanic/Latino: 58	Not Hispanic/Latino: 14	Not Hispanic/Latino: 24
Race	African American: 38	African American: 8	African American: 15
	White: 20	White: 5	White: 7
	American India or Alaskan Native: 2	American India or Alaskan Native: 1	Hawaii & Pacific Islander: 4
Major	Science: 49	Science: 9	Science: 13
	Technology: 4 Engineering:3	Technology: 1	Technology: 1
	Mathematics:4	Engineering: 1	Engineering: 5
			Mathematics: 7
	2015 (n = 22)	**2016 (n = 28)**	
Gender	Female: 14	Female: 9	
	Male: 8	Male: 19	
Ethnicity	Hispanic/Latino: 0	Hispanic/Latino: 3	
	Not Hispanic/Latino: 22	Not Hispanic/Latino: 25	
Race	African American: 4	African American: 14	
	White: 17	White: 11	
	American India or Alaskan Native: 1	Mexican: 3	
Major	Science: 11	Science: 8	
	Technology: 3	Technology: 4	
	Engineering: 3	Engineering: 13	
	Mathematics: 5	Mathematics: 3	

competency to be successful in STEM. As shown in the following Table 2, non-white students were likely to be more positive in their responses to the survey item than white students.

A Man-Whitney U test comparing non-white students' and white students' responses yielded a statistically significant result (z = 1.978, p = .048) indicating a higher rating of this survey item by non-white students.

Table 2. U. S. Space & Rocket Center: STEM competency

Race	Reported Statistics	Strongly Disagree	Disagree	Agree	Strongly Agree
White	Count (n)	2	5	11	2
	% Within Race	10.0%	25.0%	55.0%	10.0%
	% Within Responses	50.0%	50.0%	34.4%	14.3%
	% of Total	3.3%	8.3%	18.3%	3.3%
Non-White	Count (n)	2	5	21	12
	% Within Race	5.0%	12.5%	52.5%	30.0%
	% Within Responses	50.0%	50.0%	35.6%	85.7%
	% of Total	3.3%	8.3%	35.0%	20.0%

Knowledge of STEM Fields

The mean of responses to this survey item was 3.03. The "within race" percentages in the following table indicated that non-white students were more likely to agree that the field trip increased their knowledge of STEM fields. A Man-Whitney U test comparing non-white students' and white students' responses did not yield a statistically significant result.

Skills of Working on STEM Projects

This survey item specifically asked the students to rate their agreement with the statement of "The time spent in the Parachute Lab at the Space and Rocket Center in Huntsville increased my skills in working on STEM projects". The mean of the responses to this survey item was 2.88 not indicating a very positive rating of the experience of the Parachute Lab in improving STEM project skills.

The "within race" percentages shown in the Table 4 indicate that the non-white STEM students were likely to have a more positive rating of their experience of the Parachute Lab in terms of increasing their skills of working on STEM projects. A Man-Whitney U test comparing non-white students' and white students' responses was conducted. The test yielded a statistically significant result ($z = 2.409, p = .016$) indicating a higher rating of non-white undergraduate students to this survey item.

Table 3. U. S. Space & Rocket Center: Knowledge of STEM fields

Race	Reported Statistics	Strongly Disagree	Disagree	Agree	Strongly Agree
White	Count (n)	1	5	9	5
	% Within Race	5.0%	25.0%	45.0%	25.0%
	% Within Responses	25.0%	62.5%	30.0%	27.8%
	% of Total	3.3%	8.3%	18.3%	3.3%
Non-White	Count (n)	3	3	21	13
	% Within Race	7.5%	7.5%	52.5%	32.5%
	% Within Responses	75.0%	37.5%	70.0%	72.2%
	% of Total	5.0%	5.0%	35.0%	21.7%

Table 4. U. S. Space & Rocket Center: Skills of working on STEM projects

Race	Reported Statistics	Strongly Disagree	Disagree	Agree	Strongly Agree
White	Count (n)	4	6	11	3
	% Within Race	20.0%	30.0%	35.0%	15.0%
	% Within Responses	57.1%	50.0%	31.8%	15.8%
	% of Total	3.3%	8.3%	18.3%	3.3%
Non-White	Count (n)	3	6	15	16
	% Within Race	7.5%	15.0%	37.5%	40.0%
	% Within Responses	42.9%	50.0%	68.2%	84.2%
	% of Total	5.0%	10.0%	25.0%	26.7%

Persistence as STEM Majors

The mean of the students' responses to the survey item "As a result of the field trip experience, I am inclined to continue to pursue a STEM discipline" was 3.06.

The "within race" percentages shown in Table 5 reveal a similar pattern between the non-white and white STEM students in their ratings of the field trip experience as related to their inclination to continue to be a STEM major. With such a similar pattern, it was not surprising to have an insignificant result from the Mann-Whitney U test comparing non-white students' and white students' responses.

Table 5. U. S. Space & Rocket Center: Persistence as STEM majors

Race	Reported Statistics	Strongly Disagree	Disagree	Agree	Strongly Agree
White	Count (n)	2	3	8	7
	% Within Race	10.0%	15.0%	40.0%	35.0%
	% Within Responses	50.0%	33.3%	30.8%	33.3%
	% of Total	3.3%	5.0%	13.3%	11.7%
Non-White	Count (n)	2	6	18	14
	% Within Race	5.0%	15.0%	45.0%	35.0%
	% Within Responses	50.0%	66.7%	69.2%	66.7%
	% of Total	3.3%	10.0%	30.0%	23.3%

Intention to Work as STEM Professionals After Graduation

The mean of the students' responses to the survey item "As a result of the field trip experience, I am inclined to continue to pursue a STEM career track" was 3.02. The "within race" percentages shown in Table 6 below did not indicate a different pattern in the non-white and white students' responses to the survey item. A Mann-Whitney U test comparing the responses by non-white and white students did not yield a significant result.

2013 Field Trip to the Alabama Nature Center

The analysis of the 2013 field trip data specifically examined the effects of the trip on undergraduate STEM students' competency to be successful in STEM, knowledge of STEM fields, persistence as STEM majors, and intention to work as STEM professionals after graduation. In the survey data analysis, the responses were coded with "1 = strongly disagree", "2 = disagree", "3 = agree", and "4 = strongly agree".

Competency to Be Successful in STEM

The mean of the responses to the survey item "The field trip to Water World at the Alabama Nature Center increased my competency to be successful in STEM projects" was 3.64 indicating a very positive rating regarding the field trip's effects on improving students' competency to be successful in STEM.

Table 6. U. S. Space & Rocket Center: Intention to work as a STEM professional

Race	Reported Statistics	Strongly Disagree	Disagree	Agree	Strongly Agree
White	Count (n)	2	3	11	4
	% Within Race	10.0%	15.0%	55.0%	20.0%
	% Within Responses	33.3%	42.9%	40.7%	20.0%
	% of Total	3.3%	5.0%	18.3%	6.7%
Non-White	Count (n)	4	4	16	16
	% Within Race	10.0%	10.0%	40.0%	40.0%
	% Within Responses	66.7%	57.1%	59.3%	80.0%
	% of Total	6.7%	6.7%	26.7%	26.7%

Table 7. Alabama Nature Center: Competency to be successful in STEM

Race	Reported Statistics	Disagree	Agree	Strongly Agree
White	Count (n)	0	1	4
	% Within Race	0.0%	20.0%	80.0%
	% Within Responses	0.0%	33.3%	40.0%
	% of Total	0.0%	7.1%	28.6%
Non-White	Count (n)	1	2	6
	% Within Race	11.1%	22.2%	66.7%
	% Within Responses	10.0%	66.7%	60.0%
	% of Total	7.1%	14.3%	42.9%

Due to small sample size of the respondents to this survey, the "within race" and "within responses" percentages shown in Table 7 might be misleading. Therefore, the means of non-white students' and white students' responses were calculated. The means for non-white students' and white students' responses were 3.56 and 3.80 respectively. A Mann-Whitney U test was conducted, and the test result indicated an insignificant difference between non-white and white students' responses regarding the effects of the field trip experience on their competency to be successful in STEM.

Knowledge of STEM Fields

The mean of the responses to the survey item "The field trip to Water World at the Alabama Nature Center strengthened my knowledge of STEM fields" was 3.86 indicating a very positive rating of the field trip's effects on strengthening knowledge of STEM fields. As shown in the following Figure 2, no students rated "strongly disagree" or "disagree."

The means for non-white students' and white students' responses were 3.78 and 4.00 respectively. The mean difference was not statistically significant based on the results of a Mann-Whitney U test.

Persistence as STEM Majors

The mean of the responses to the survey item "As a result of attending the Project Engage field experience at Water World, I am inclined to continue to pursue a STEM discipline" was 3.79 indicating a very high rating regarding the field trip's effects on enhancing the students' persistence as STEM majors.

As shown in Figure 3, no students rated "strongly disagree" or "disagree." The means for non-white students' and white students' responses were 3.78 and 3.80. A Mann-Whitney U test was conducted to compare non-white students' and white students' responses and the test results did not indicate significant difference.

Intention to Work as STEM Professionals After Graduation

The mean of the responses to the survey item "As a result of attending the Project Engage field experience at Water World, I am inclined to continue to pursue a STEM career track" was 3.79 indicating a very positive opinion about the field trip's effects to strengthen STEM undergraduate students'

Figure 2. Alabama Nature Center: Knowledge of STEM fields

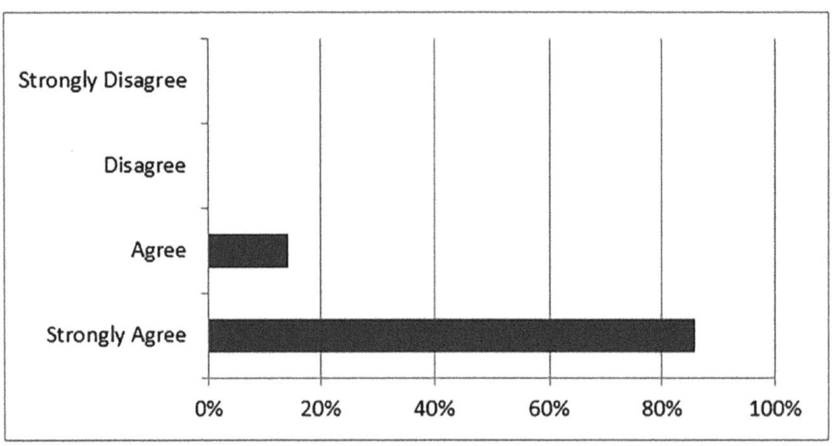

Figure 3. Alabama Nature Center: Persistence as STEM Majors

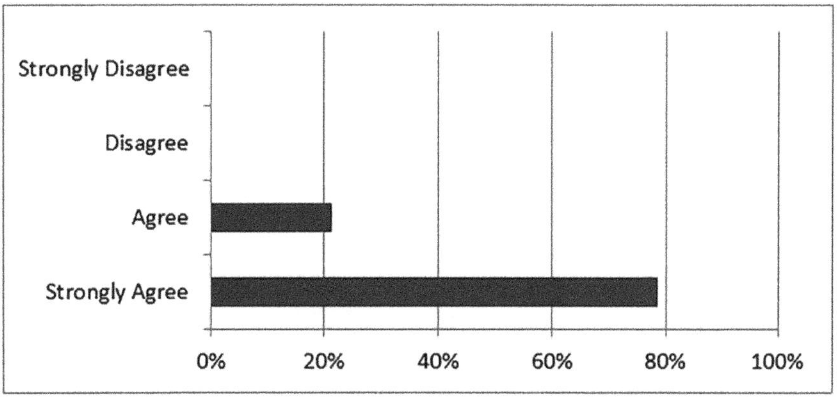

intention to pursue a STEM career. As shown in the following Figure 4, no students rated "strongly disagree" or "disagree."

The means for non-white students' and white students' responses were 3.67 and 4.00. A Mann-Whitney U test was conducted to compare non-white students' and white students' responses, and the test results did not indicate a significant difference.

Discovery Based Activities

The 2013 field trip to the Alabama Nature Center included five discovery based activities:

Figure 4. Alabama Nature Center: Intention to work as a STEM professional

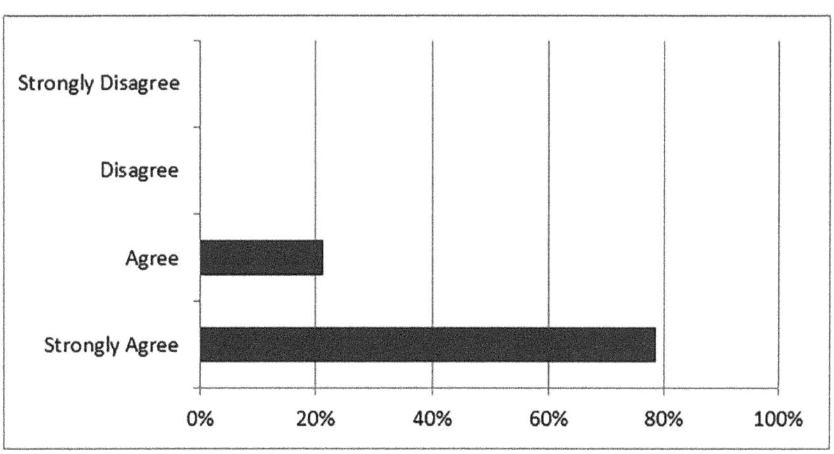

1. Collecting insect larvae from a creek (Aquatic Life Roundup) (Creek Culture)
2. Observing/handling the "shocked" fish (Water Quality)
3. Fishing
4. Discussion of watersheds with a trip to the stream (Wetland Hike)
5. Observation and/or handling of plants and wildlife such as snakes (Alabama Animals)

The 2013 field trip survey also asked the students to respond to the survey items of "Which activity increased your STEM knowledge the most?" and "Which activity increased your STEM skills the most?" The frequency distribution of the students' responses are shown below in Table 8.

Mercedes-Benz U.S. International Incorporated

The analysis of the 2016 field trip data specifically examined the effects of the trip on undergraduate STEM students' competency to be successful in STEM, knowledge of STEM fields, knowledge of real world STEM applications, persistence as STEM majors, and intention to work as STEM professionals after graduation. In the survey data analysis, the responses were coded with "1 = strongly disagree", "2 = disagree", "3 = agree", and "4 = strongly agree".

Competency to Be Successful in STEM

The mean of the responses to the survey item "The field trip to Mercedes Benz increased my competency to be successful in working on STEM projects" was 3.39 indicating a positive rating regarding the field trip's effects on improving students' competency to be successful in STEM. The "within race" percentages in the following Table 9 show that the white students were less positive than non-white students about the effects of the field trip on their competency to be successful in STEM.

Table 8. Alabama Nature Center: Discovery based activities

Question	Activity 1	Activity 2	Activity 3	Activity 4	Activity 5
Which activity increased your STEM knowledge the most?	4 (29%)	4 (29%)	0 (0%)	4 (29%)	2 (14%)
Which activity increased your STEM skills the most?	7 (50%)	4 (29%)	1 (7%)	1 (7%)	1 (7%)

Table 9. Mercedes-Benz: Competency to be Successful in STEM

Race	Reported Statistics	Disagree	Agree	Strongly Agree
White	Count (n)	3	3	5
	% Within Race	27.3%	27.3%	45.5%
	% Within Responses	75.0%	33.3%	33.3%
	% of Total	10.7%	10.7%	17.9%
Non-White	Count (n)	1	6	10
	% Within Race	5.9%	35.3%	58.8%
	% Within Responses	25.0%	66.7%	66.7%
	% of Total	3.6%	21.4%	35.7%

A Mann-Whitney U test was conducted and the test result indicated insignificant difference between non-white and white students' responses regarding the effects of field trip experience on their competency to be successful in STEM.

Knowledge of STEM Fields

The mean of the responses to the survey item "The field trip to Mercedes Benz improved my knowledge of STEM fields" was 3.43 indicating a positive rating of the Mercedes Benz field trip's effects on increasing knowledge of STEM fields. The response frequency distribution by the non-white and white students is shown in Table 10.

The "within race" percentages in the above table indicate a rather obvious pattern that the non-white students were more positive about the effects of the field trip on their knowledge of STEM fields. The white and non-white students' responses were compared using a Mann-Whitney U test and the result did not indicate a significant difference.

Persistence as STEM Majors

The mean of the responses to the survey item "As a result of attending the Project Engage field experience at Mercedes Benz, I am inclined to continue to pursue a STEM discipline" was 3.43 indicating a positive rating of the Mercedes Benz field trip's effects on increasing the students' intention to stay as STEM majors. The response frequency distribution by the non-white and white students is shown below in Table 11.

Table 10. Mercedes-Benz: Knowledge of STEM fields

Race	Reported Statistics	Disagree	Agree	Strongly Agree
White	Count (n)	2	4	5
	% Within Race	18.2%	36.4%	45.5%
	% Within Responses	100.0%	30.8%	38.5%
	% of Total	7.1%	14.3%	17.9%
Non-White	Count (n)	0	9	8
	% Within Race	0.0%	52.9%	47.1%
	% Within Responses	0.0%	69.2%	61.5%
	% of Total	0.0%	32.1%	28.6%

Table 11. Mercedes-Benz: Persistence as STEM majors

Race	Reported Statistics	Disagree	Agree	Strongly Agree
White	Count (n)	2	3	6
	% Within Race	18.2%	27.3%	54.5%
	% Within Responses	100.0%	25.0%	42.9%
	% of Total	7.1%	10.7%	21.4%
Non-White	Count (n)	0	9	8
	% Within Race	0.0%	52.9%	47.1%
	% Within Responses	0.0%	75.0%	57.1%
	% of Total	0.0%	32.1%	28.6%

The "within race" percentages in the above frequency distribution table indicate that the non-white students were more positive in their response to this survey item than the white students. A Mann-Whitney U test was conducted to compare the white and non-white students' responses, and the result did not indicate a significant difference.

Intention to Work as STEM Professionals After Graduation

The mean of the responses to the survey item "As a result of attending the Project Engage field experience at Mercedes Benz, I am inclined to continue to pursue a STEM career track" was 3.39 indicating a positive rating of the Mercedes Benz field trip's effects on increasing the students' intention to pursue a STEM career. The response frequency distribution by the non-white and white students is shown below in Table 12.

Table 12. Mercedes-Benz: Intention to work as a STEM professional

Race	Reported Statistics	Disagree	Agree	Strongly Agree
White	Count (n)	2	3	6
	% Within Race	18.2%	27.3%	54.5%
	% Within Responses	66.7%	27.3%	42.9%
	% of Total	7.1%	10.7%	21.4%
Non-White	Count (n)	1	8	8
	% Within Race	5.9%	47.1%	47.1%
	% Within Responses	33.3%	72.7%	57.1%
	% of Total	3.6%	28.6%	28.6%

The "within race" percentages in the above table do not indicate a clear pattern regarding white and non-white students' responses to this survey item. A Mann-Whitney U test was conducted to compare the white and non-white students' responses, and the result did not indicate a significant difference.

Knowledge of Real World STEM Applications

The survey of the field trip to Mercedes Benz also included an item regarding knowledge of real world STEM applications: "The field trip to Mercedes Benz improved my knowledge and understanding of real world applications of STEM". The mean of the responses to this survey item was 3.46 indicating a positive rating of the field trip's effects on increasing knowledge of real world STEM applications.

The "within race" percentages in the above response frequency distribution table (Table 13) show that non-white students had a more positive rating of the Mercedes Benz field trip in increasing their knowledge of real world STEM applications. The white and non-white students' responses were analyzed using a Mann-Whitney U test, and the test results did not indicate a significant difference.

STEM Challenge Surveys

The Crime Scene Investigation STEM Challenge was conducted among 2014 and 2015 Cohorts. The survey data collected from the two Cohorts were analyzed together. The demographic information for the survey participants is shown in Table 14. The instrument used to collect STEM Challenge data consisted of demographics questions about gender, race, ethnicity, and

Table 13. Mercedes-Benz: Knowledge of real world STEM applications

Race	Reported Statistics	Disagree	Agree	Strongly Agree
White	Count (n)	2	4	5
	% Within Race	18.2%	36.4%	45.5%
	% Within Responses	100.0%	36.4%	33.3%
	% of Total	7.1%	14.3%	17.9%
Non-White	Count (n)	0	7	10
	% Within Race	0.0%	41.2%	58.8%
	% Within Responses	0.0%	63.6%	66.7%
	% of Total	0.0%	25.0%	35.7%

STEM majors and survey items regarding the effects of the field trips on the undergraduate students' interest in STEM fields, knowledge of STEM careers, persistence as STEM majors, and intention to work as STEM professionals after graduation. The survey questions were positively worded to avoid confusion for respondents (Netemeyer, Bearden, & Sharma, 2003) and cognitive burdens for information processing (Dillman, 2006). The survey items adopted a 4 point Likert-like scale (i.e., Strongly disagree, disagree, agree, and strongly agree). The 2014 and 2015 Cohorts' data were analyzed together and the responses were coded with "1 = strongly disagree", "2 = disagree", "3 = agree", and "4 = strongly agree".

Table 14. Demographics of STEM Challenge survey participants

	2014 (n = 13)	2015 (n = 35)
Gender	Female: 9	Female: 17
	Male: 4	Male: 18
Ethnicity	Hispanic/Latino: 0	Hispanic/Latino: 0
	Not Hispanic/Latino: 13	Not Hispanic/Latino: 35
Race	African American: 7	African American: 12
	White: 5	White: 21
	American India or Alaskan Native: 1	Hawaii/Pacific Islander: 2
Major	Science: 10	Science: 18
	Technology: 1	Technology: 4
	Engineering: 1	Engineering: 9
	Mathematics: 1	Mathematics: 4

Interest in STEM Fields

The mean of the responses to the survey item "The STEM Challenge strengthened my interest in STEM fields through the Crime Scene Investigation activity" was 3.17 indicating the students' positive rating of the STEM Challenge in increasing their interest in STEM fields. The response frequency distributions of the white and non-white students' responses are presented in the Table 15.

The "within race" percentages indicate that the non-white students had more positive opinions about the effects of the STEM Challenge on increasing their interest in STEM fields. The Mann-Whitney U test for comparing non-white and white students' responses did not yield significant results.

Knowledge of STEM Fields

The mean to the survey item "The STEM Challenge increased my knowledge of STEM fields through the Crime Scene Investigation activity" was 3.25 indicating a positive rating about the effects of the STEM Challenge on increasing knowledge of STEM fields.

As shown by the above response frequency distribution table (Table 16), the non-white and white students demonstrated similar patterns in their rating of the survey statement with equal "within responses" percentages for "strongly disagree", "disagree", and "agree". The Mann-Whitney U test for comparing non-white and white students' responses did not yield significant results.

Table 15. STEM Challenge: Interest in STEM fields

Race	Reported Statistics	Strongly Disagree	Disagree	Agree	Strongly Agree
White	Count (n)	2	3	10	11
	% Within Race	7.7%	11.5%	38.5%	42.3%
	% Within Responses	100.0%	50.0%	45.5%	61.1%
	% of Total	4.2%	6.3%	20.8%	22.9%
Non-White	Count (n)	0	3	12	7
	% Within Race	0.0%	13.6%	54.5%	31.8%
	% Within Responses	0.0%	50.0%	54.5%	38.9%
	% of Total	0.0%	6.3%	25.0%	14.6%

Table 16. STEM Challenge: Knowledge of STEM fields

Race	Reported Statistics	Strongly Disagree	Disagree	Agree	Strongly Agree
White	Count (n)	1	3	9	13
	% Within Race	3.8%	11.5%	34.6%	50.0%
	% Within Responses	50.0%	50.0%	50.0%	59.1%
	% of Total	2.1%	6.3%	18.8%	27.1%
Non-White	Count (n)	1	3	9	9
	% Within Race	4.5%	13.6%	40.9%	40.9%
	% Within Responses	50.0%	50.0%	50.0%	40.9%
	% of Total	2.1%	6.3%	18.8%	18.8%

Persistence as STEM Majors

The mean to the survey item "As a result of participating in the STEM Challenge, I am inclined to continue to pursue a STEM discipline as a major in college" was 3.23 indicating a positive rating about the effects of the STEM Challenge on increasing the students' persistence as STEM majors.

The "within race" percentages in the above response frequency distribution table show that the non-white students, as compared with white students, were less positive about the effects of the STEM Challenge on increasing their persistence as STEM majors (see Table 17). A Mann-Whitney U test was conducted to compare the non-white and white students' response to the survey item, but the test results did not indicate a significant difference.

Table 17. STEM Challenge: Persistence as STEM majors

Race	Reported Statistics	Disagree	Agree	Strongly Agree
White	Count (n)	3	12	11
	% Within Race	11.5%	46.2%	42.3%
	% Within Responses	33.3%	63.2%	55.0%
	% of Total	6.3%	25.0%	22.9%
Non-White	Count (n)	6	7	9
	% Within Race	27.3%	31.8%	40.9%
	% Within Responses	66.7%	36.8%	45.0%
	% of Total	12.5%	14.6%	18.8%

Intention to Work as STEM Professionals After Graduation

The mean to the survey item "As a result of participating in the STEM Challenge, I am inclined to continue to pursue a STEM career after college" was 3.33 indicating a positive rating about the effects of the STEM Challenge on strengthening the students' intention to become STEM professionals after graduation.

The "within race" percentages in the above response frequency distribution indicate that the non-white students and the white students showed a similar pattern in their responses to the survey item. A Mann-Whitney U test was conducted to compare the white and non-white students' responses and no significant difference was identified.

Invention Convention Survey

The 2012 Cohort participated in the Lego Robotics Invention Convention, and Mr. Todd May, former project director for NASA and current director at the Marshall Space Flight Center was invited as keynote speaker for this event. Survey data were collected from the participants. The demographic information of the students who responded to the survey is shown in Table 18. The instrument used to collect data from the 2012 Cohort regarding this Invention Convention consisted of demographics questions about gender, race, ethnicity, and STEM majors and survey items regarding the effects of the field trips on the undergraduate students' interest in STEM fields, knowledge of STEM fields, persistence as STEM majors, and intention to work as STEM professionals after graduation. The survey items were positively worded to avoid confusion for respondents (Netemeyer, Bearden, & Sharma, 2003) and

Table 18. STEM Challenge: Intention to work as a STEM professional

Race	Reported Statistics	Disagree	Agree	Strongly Agree
White	Count (n)	2	13	11
	% Within Race	7.7%	50.0%	42.3%
	% Within Responses	50.0%	54.2%	55.0%
	% of Total	4.2%	27.1%	22.9%
Non-White	Count (n)	2	11	9
	% Within Race	9.1%	50.0%	40.9%
	% Within Responses	50.0%	45.8%	45.0%
	% of Total	4.2%	22.9%	18.8%

cognitive burdens for information processing (Dillman, 2006). The survey items adopted a 4 point Likert-like scale (i.e., Strongly disagree, disagree, agree, and strongly agree). The responses to the survey items were coded with "1 = strongly disagree", "2 = disagree", "3 = agree", and "4 = strongly agree".

Interest in STEM Fields

One of the survey items asked the students to rate the overall effects of the Invention Convention on their interest in STEM fields: "Overall, the Invention Convention competition strengthened my interest in STEM fields". The mean of the responses to this survey item was 2.94 not indicating a very positive rating regard the Invention Convention's effect on increasing the interest in STEM fields. The response frequency distribution of the non-white and white students is presented in the Table 19.

The "within race" percentages in the above table indicate that the white and non-white students did not show obvious difference in the frequency distribution of their responses. A Mann-Whitney U test was conducted to compare the white and non-white students' responses, and no significant difference was identified.

One of the Invention Convention survey items specifically focused on the effects of the keynote speaker on the students' interest in STEM fields: "The information provided by the keynote speaker strengthened my interest in the STEM fields." The mean of responses to this survey item was 3.16 indicating an overall positive rating for this survey item. The frequency distribution of the white and non-white student's responses is presented in Table 20.

Table 18. Demographic information of the Lego Robotics Invention Convention survey participants

Gender	Female: 25
	Male: 8
Ethnicity	Hispanic/Latino: 0
	Not Hispanic/Latino: 33
Race	African American: 17
	White: 16
Major	Science: 24
	Technology: 3
	Engineering: 2
	Mathematics: 4

Table 19. Lego Robotics Invention Convention: Interest in STEM fields

Race	Reported Statistics	Strongly Disagree	Disagree	Agree	Strongly Agree
White	Count (n)	1	3	10	2
	% Within Race	6.3%	18.8%	62.5%	12.5%
	% Within Responses	100.0%	50.0%	50.0%	33.3%
	% of Total	2.1%	6.3%	18.8%	27.1%
Non-White	Count (n)	0	3	10	4
	% Within Race	0.0%	17.6%	58.8%	23.5%
	% Within Responses	0.0%	50.0%	50.0%	66.7%
	% of Total	0.0%	9.1%	30.3%	12.1%

Table 20. Lego Robotics Invention Convention: Keynote speakers interest in STEM

Race	Reported Statistics	Strongly Disagree	Disagree	Agree	Strongly Agree
White	Count (n)	1	1	11	3
	% Within Race	6.3%	6.3%	68.8%	18.8%
	% Within Responses	100.0%	50.0%	52.4%	33.3%
	% of Total	3.0%	3.0%	33.3%	9.1%
Non-White	Count (n)	0	3	10	4
	% Within Race	0.0%	5.9%	58.8%	23.5%
	% Within Responses	0.0%	50.0%	47.6%	66.7%
	% of Total	0.0%	3.0%	30.3%	18.2%

A comparison of "within race" percentages for the white and non-white students did not identify obvious difference between the two group of students' responses. The results from the Mann-Whitney U test did not yield significant difference in white and non-white students' responses.

Knowledge in STEM Fields

One of the survey items asked the students to rate the overall effects of the Invention Convention on their knowledge in STEM fields: "Overall the Invention Convention increased my knowledge of STEM fields". The mean of the responses to this survey item was 3.00 not indicating a very positive rating in regard to the Invention Convention's effect on increasing the students' knowledge of STEM fields. The response frequency distribution of the non-white and white students is presented in the Table 21.

Table 21. Lego Robotics Invention Convention: Knowledge in STEM Fields

Race	Reported Statistics	Strongly Disagree	Disagree	Agree	Strongly Agree
White	Count (n)	2	1	11	2
	% Within Race	12.5%	6.3%	68.8%	12.5%
	% Within Responses	100.0%	25.0%	57.9%	25.0%
	% of Total	6.1%	3.0%	33.3%	6.1%
Non-White	Count (n)	0	3	8	6
	% Within Race	0.0%	17.6%	47.1%	35.3%
	% Within Responses	0.0%	75.0%	42.1%	75.0%
	% of Total	0.0%	9.1%	24.2%	18.2%

The "within race" percentages in the above table indicate that the non-white students tended to have a higher rating of this survey item with a higher percentage of them in the "strongly agree" category. The mean of non-white students' responses was 3.17 and that of white students' response was 2.81. A Mann-Whitney U test was conducted to compare the white and non-white students' responses, and no significant difference was identified.

Another survey item related to knowledge in STEM fields asked the students to rate the statement: "The information provided by the keynote speaker increased my knowledge in the STEM fields." The mean of the responses to this survey item was 3.12 indicating the students' positive opinion about the keynote speaker's effects on increasing the knowledge in STEM fields. The response frequency distribution of the non-white and white students is presented in Table 22.

Table 22. Lego Robotics Invention Convention: Keynote speaker increased knowledge

Race	Reported Statistics	Disagree	Agree	Strongly Agree
White	Count (n)	1	13	2
	% Within Race	6.3%	81.3%	12.5%
	% Within Responses	50.0%	59.1%	22.2%
	% of Total	3.0%	39.4%	6.1%
Non-White	Count (n)	0	9	7
	% Within Race	5.9%	52.9%	41.2%
	% Within Responses	50.0%	40.9%	77.8%
	% of Total	3.0%	27.3%	21.2%

As indicated by the "within race" percentages in the above table, a higher percentage of non-white students were in the "strongly agree" category than white students. The mean of non-white students' responses was 3.35 and that of white students' response was 3.06. A Mann-Whitney U test was conducted to compare the white and non-white students' responses, and no significant difference was identified.

Persistence as STEM Majors

The mean to the survey item "As a result of participating in the Invention Convention, I am inclined to continue to pursue a STEM discipline as a major in college" was 3.09 indicating a positive rating of the Invention Convention in terms of its effect on enhancing the students' persistence as STEM majors. The response frequency distribution of the non-white and white students is presented in Table 23.

The "within race" percentages in the above table indicate that a higher percent of non-white students than white students were in the "strongly agree" category. The mean of non-white students' responses was 3.41 and that of white students' responses was 2.75. A Mann-Whitney U test was conducted to compare non-white and white students' responses, and the test result yielded a significant difference ($z = 2.308$, $p = .021$).

Intention to Work as STEM Professionals After Graduation

The mean to the survey item "As a result of participating in the Invention Convention, I am inclined to continue to pursue a STEM career after college" was 3.06 indicating a positive rating of the Invention Convention in terms of

Table 23. Lego Robotics Invention Convention: Persistence as STEM majors

Race	Reported Statistics	Strongly Disagree	Disagree	Agree	Strongly Agree
White	Count (n)	2	2	10	2
	% Within Race	12.5%	12.5%	62.5%	12.5%
	% Within Responses	100.0%	50.0%	62.5%	18.2%
	% of Total	6.1%	6.1%	30.3%	6.1%
Non-White	Count (n)	0	2	6	9
	% Within Race	0.0%	11.8%	35.3%	52.9%
	% Within Responses	0.0%	50.0%	37.5%	81.8%
	% of Total	0.0%	6.1%	18.2%	27.3%

its effect on increasing the students' intention to become STEM professionals after graduation. The frequency distribution of the non-white and white students' responses is presented in Table 24.

The "within race" percentages in the above table indicate that a higher percent of non-white students than white students were in the "strongly agree" category. The mean of non-white students' responses was 3.35 and that of white students' responses was 2.75. A Mann-Whitney U test was conducted to compare non-white and white students' responses, and the test result yielded a significant difference ($z = 2.150, p = .032$).

Team Building Alabama 4-H Center Survey

The old days of single-inventor innovations have long been replaced with team work and team research in STEM fields. Teamwork skill is one of the must-have soft skills for STEM professionals. To prepare STEM undergraduate students as tomorrow' STEM professional, Project Engage designed and implemented teambuilding themed field trips and in class team building instructional activities.

The 2014 and 2015 field trips to the Alabama 4-H Center were team building field trips. The survey data collected from the two field trips were analyzed together to assess the effects of the activities on team cooperation and self-confidence. The demographic information of the 2014 and 2015 field trip survey participants is shown in Table 1. The teambuilding activities during the 4-H Center field trips included the Giant Rope Swing activity, the Clover Hop activity, and the Rock Wall activity.

Table 24. Lego Robotics Invention Convention: Intentional to work as a STEM professional

Race	Reported Statistics	Strongly Disagree	Disagree	Agree	Strongly Agree
White	Count (n)	2	2	10	2
	% Within Race	12.5%	12.5%	62.5%	12.5%
	% Within Responses	100.0%	66.7%	52.6%	22.2%
	% of Total	6.1%	6.1%	30.3%	6.1%
Non-White	Count (n)	0	1	9	7
	% Within Race	0.0%	5.9%	52.9%	41.2%
	% Within Responses	0.0%	33.3%	47.4%	77.8%
	% of Total	0.0%	3.0%	27.3%	21.2%

Team Cooperation

Three items in the 4-H Center survey were related to team cooperation: (1) The Giant Rope Swing activity helped build team cooperation; (2) The Clover Hop activity helped build team cooperation; and (3) The Rock Wall activity helped build team cooperation. The means of the responses to the three survey items were 3.60, 3.69, and 2.98 indicating that, compared to the Rock Wall activity, the Giant Rope Swing activity and the Clover Hop activity were better received by the students in helping build team cooperation. The response frequency distribution for the three activities in terms of building team cooperation is shown in Table 25.

Personal Confidence

Three items in the 4-H Center were related to personal confidence: (1) The Giant Rope Swing activity helped build my personal confidence; (2) The Clover Hop activity helped build my personal confidence; and (3) The Rock Wall activity helped build my personal confidence. The means of the responses to the three survey items were 3.69, 3.17, and 3.33 indicating that all three activities were good at helping build the students' personal confidence. The response frequency distribution for the three activities in terms of building personal confidence is shown in Table 26.

Team Building Class Instruction Survey

In addition to the teambuilding activities during the 4-H Center field trip, Project Engage also brought in an outside consultant with expertise in teambuilding

Table 25. Building team cooperation: Response frequency distribution

	Giant Rope Swing Activity		Clover Hop Activity		Rock Wall Activity	
	Frequency	Relative Frequency	Frequency	Relative Frequency	Frequency	Relative Frequency
Strongly Disagree	1	2%	0	0%	5	10%
Disagree	4	8%	4	8%	8	17%
Agree	8	17%	7	15%	18	38%
Strongly Agree	35	73%	37	77%	17	35%

Table 26. Building personal confidence: Response frequency distribution

	Giant Rope Swing Activity		Clover Hop Activity		Rock Wall Activity	
	Frequency	Relative Frequency	Frequency	Relative Frequency	Frequency	Relative Frequency
Strongly Disagree	1	2%	5	10%	2	4%
Disagree	2	4%	6	13%	2	4%
Agree	8	17%	13	27%	22	46%
Strongly Agree	37	77%	24	50%	22	46%

to lead each freshman STEM 101 class in an interactive exchange in what it takes to make an effective team. A key element in the training was creating team contracts. The survey data of the teambuilding session of freshman STEM 101 were available from 56 freshman students. The survey items adopted a 4 point Likert-like scale (i.e., Strongly disagree, disagree, agree, and strongly agree). The following items in the survey were analyzed, and the responses were coded with "1 = strongly disagree", "2 = disagree", "3 = agree", and "4 = strongly agree".

The Team Effectiveness Strategies

- **Item:** The Team Effectiveness Strategies discussion increased my competency to be successful in working on STEM projects.

The mean to this survey item was 3.62 indicating a very positive rating of the Team Effective Strategies discussion in increasing the students' competency to be successful in STEM projects. The frequency distribution of the students' responses is shown in Table 27.

- **Item:** The Team Effectiveness Strategies session increased my knowledge and interest in STEM fields.

The mean to this survey item was 3.39 indicating a positive rating of the Team Effective Strategies discussion in increasing the students' knowledge and interest in STEM fields. The frequency distribution of the students' responses is shown in the table below:

Table 27. Team effectiveness strategies: Competency to work on projects

	Frequency	Relative Frequency
Strongly Disagree	2	3.5%
Disagree	0	0%
Agree	22	38.5%
Strongly Agree	33	58%

Table 28. Team effectiveness strategies: Knowledge and interest in STEM

	Frequency	Relative Frequency
Strongly Disagree	3	5%
Disagree	2	3.5%
Agree	22	38.5%
Strongly Agree	30	53%

- **Item:** I am more aware of Team Effectiveness ideas than before the Team Effectiveness discussions.

The mean to this survey item was 3.53 indicating a very positive rating of the Team Effective Strategies discussion in improving the students' team effective ideas. The frequency distribution of the students' responses is shown in Table 29.

- **Item:** Developing the team contract assignment was helpful to me as a team member.

The mean to this survey item was 3.54 indicating a very positive rating of the helpfulness of the contract assignment. The frequency distribution of the students' responses is shown in Table 30.

Table 29. Team effectiveness strategies: Team effectiveness ideas

	Frequency	Relative Frequency
Strongly Disagree	2	3.5%
Disagree	2	3.5%
Agree	17	30%
Strongly Agree	36	63%

Table 30. Team contract was helpful to me as a team member

	Frequency	Relative Frequency
Strongly Disagree	2	3.5%
Disagree	2	3.5%
Agree	16	28%
Strongly Agree	37	65%

SUMMARY AND FUTURE RESEARCH DIRECTIONS

Field Trips

The analyses of the data collected from the 2012 field trip to the Space and Rocket Center, 2013 field trip to the Alabama Nature Center, and 2016 field trip to Mercedes Benz focused on assessing the effects of the field trips on improving competency to be successful in STEM, knowledge of STEM fields, persistence as STEM majors, and intention to work as STEM professionals after graduation. The means for the responses to these four survey items are summarized in the following table:

As indicated in Table 31, the means for the 2012 and 2016 field trips were lower than those for the 2013 field trips. Research has indicated that hands-on learning activities provide students with more realistic and exciting experiences (Franklin & Peat 2005) and lead to positive motivational outcomes (Thompson & Soyibo, 2002). Since the 2013 field trip to the Alabama Nature Center involved five discovery based activities (i.e., collecting Insect Larvae from a creek (Aquatic Life Roundup), observing/handling the "shocked" fish

Table 31. Summary of field trip survey response means

	Means			
	Competency to be successful in STEM	Knowledge of STEM fields	Persistence as STEM majors	Intention to work as STEM professionals
2012 field trip to the Space and Rocket Center	2.93	3.03	3.06	3.02
2013 field trip to the Alabama Nature Center	3.64	3.86	3.79	3.79
2016 field trip to Mercedes Benz	3.39	3.43	3.43	3.39

(Water Quality), fishing, discussion of Watersheds with a trip to the stream (Wetlands), and observation and/or handling of plants and wildlife such as snakes (Alabama Wildlife)), those activities may explain why the students of the 2013 field trip rated more positively to the four survey items. This result indicates that future research is needed to confirm if hands-on activity integrated field trips can better improve undergraduate STEM students' competency to be successful in STEM, knowledge of STEM fields, persistence as STEM majors, and intention to work as STEM professionals after graduation.

Although the trend that the non-white students' responses were more positive than those of the white students regarding the impacts regarding the field trips was identified throughout the 2012, 2013, and 2016 field trip data, only two Mann-Whitney U tests from the 2012 field trip data identified a significant difference between non-white students and white students in their responses related to "competency to be successful in STEM" and "skills of working on STEM projects." Future research is needed to reveal if the above mentioned trend can be concluded among non-white and white undergraduate STEM students in general.

Competitive Events

Stem Challenge

The analyses of the data collected from the 2014 and 2015 STEM Challenge focused on assessing the effects of the STEM Challenge events on improving interest in STEM fields, knowledge of STEM fields, persistence as STEM majors, and intention to work as STEM professionals after graduation. The respective means of the responses to these four survey items 3.17, 3.25, 3.23, and 3.33 indicate the students' positive rating of the two STEM Challenge events. Although the trend that non-white students were more positive than white students about the effects of the STEM Challenge events in terms of the above mentioned four aspects, all four Mann-Whitney U tests failed to conclude that this difference is significant among non-white and white undergraduate STEM students in general. Future research is needed to further investigate if such difference exists among non-white and white students in similar STEM Challenge events.

Robotics Invention Convention

The 2012 Lego Robotics Invention Convention survey included survey items assessing the Lego Robotics Invention Convention's effects on increasing interest in STEM fields, knowledge in STEM fields, persistence as STEM majors, and intention to work as STEM professionals after graduation. The means of the responses to these four survey items were 2.94, 3.00, 3.09, and 3.06, not indicating a very positive rating by the students, especially regarding increasing interest in STEM fields. Two Mann Whitney U tests indicates that the Invention Convention has significantly more positive effects on non-white students in terms of their persistence as STEM majors and intention to work as STEM professional after graduation. No information is available to explain why this is the case. Research endeavors intended to reveal such information will be very helpful in strengthening and diversifying the STEM pipeline.

Team Building

The 2014 and 2015 field trips to the Alabama 4-H Center were field trips with teambuilding as the main theme. The means of the survey responses regarding the effects of three teambuilding activities (i.e., the Giant Rope Swing activity, the Clover Hop activity, and the Rock Wall activity) on team cooperation were 3.60, 3.69, and 2.98 respectively, and the means of the survey responses regarding the effects of the three teambuilding activities on personal confidence were 3.69, 3.17, and 3.33 respectively, indicating the overall positive effects of the three teambuilding activities on the students' team cooperation and personal confidence. In addition to the field trips to the Alabama 4-H Center, the teambuilding session in the freshman 101 class also received positive ratings from the students regarding its effects on increasing competency to be successful in STEM projects ($M = 3.62$), knowledge and interest in STEM fields ($M = 3.39$), team effectiveness ideas ($M = 3.53$), and ability as a team member ($M = 3.54$). The result that the Rock Wall activity received less positive rating from the students regarding its effect on team cooperation calls for future research on how characteristics of teambuilding activities are related to their effects of improving teamwork skills.

CONCLUSION

The proposed Three Pillar Approach for Preparing Tomorrow's STEM Professionals includes developing: 1) knowledge of STEM careers and professionals, 2) abilities to work in STEM fields, and 3) ways of working as STEM professionals. Project Engage establishes these three pillars through STEM oriented career exploration field trips, competitive events, and guest speakers. These activities address specific pillars of the model, yet there are overlapping components. First, "knowledge of STEM careers and professionals" is primarily addressed through the STEM career exploration field trips. Engage participants interact with STEM professionals in their work environments gaining first-hand knowledge of STEM professionals and careers. However, invited STEM speakers also introduce the students to STEM professionals who, because of careful selection by the Project Team, serve as role models for the students. Secondly, STEM competitive events assist students in developing the abilities to work in STEM fields such as social and communication skills. Through competition, teams must work collaboratively acquiring social skills such as teamwork and leadership. Project Engage competitions also include a communications component, so as to cultivate written and oral communication skills. Furthermore, STEM competitive events are key to introducing students to the "ways of working" as a STEM professional. Teams must work together collaboratively to solve "crimes" using interdisciplinary knowledge and skills, design and test objects based on specifications provided, build and program robots, and much more. These challenges require critical thinking and problem-solving skills that STEM professionals use in their careers. The three pillar approach is a comprehensive methodology for preparing high school and college students, specifically underrepresented groups from rural locales, to be the STEM professionals of tomorrow. Survey results from the activities reveal a trend of non-white students rating the events higher than white students. Although typically, the results of testing for statistical differences revealed few statistically significant differences, the consistent trend of higher ratings by non-white students warrants further investigation.

ACKNOWLEDGMENT

This research was supported by the U. S. Department of Education Minority Science and Engineering Program [grant number P120A120099] and the U. S. Department of Education Minority Science and Engineering Program [grant number P120A160122].

REFERENCES

Alabama 4 H Center. (2017). *Team building at the Alabama 4-H Center.* Retrieved from http://www.aces.edu/4-H-youth/4H-Center/facilities/teamBuilding/index.php#ClimbingWall

Alabama Wildlife Federation. (2017). *Lanark Field Days Program.* Retrieved from https://www.alabamawildlife.org/lanark-field-day-program-theme-options

Allan, A. (2017). *Boosting learning by combining teamwork and STEM.* National Afterschool Association. Retrieved from https://naaweb.org/professional-development/item/718-how-to-boost-learning-by-combining-teamwork-and-stem

Bandura, A. (1986). *Social foundations of thought and action: A social cognitive theory.* Englewood Cliffs, NJ: Prentice-Hall.

Bandura, A. (1988). Organizational Application of Social Cognitive Theory. *Australian Journal of Management, 13*(2), 275–302. doi:10.1177/031289628801300210

Baynes, D. L. (2003). The Minority Aviation Education Association, Inc. *Science.* Retrieved from http://www.sciencemag.org/careers/2003/08/minority-aviation-education-association-inc.kantis

Caglak, S. (2017). Does hands-on science practices make an ımpact on achievement in science? A meta-analysis. *Journal of Education in Science Environmental Health, 3*(1), 69–87. PMID:28637469

Change the Equation & AMGEN Foundation. (2016). *Students on STEM: More hands-on, real world experiences.* Author.

Courtright, S. H., McCormick, B. W., Mistry, S., & Wang, J. (2017). Quality charters or quality members? A control theory perspective on team charters and team performance. *The Journal of Applied Psychology*, *102*(10), 1462–1470. doi:10.1037/apl0000229 PMID:28530414

DiLisi, G. A., Eppell, S. A., & Upton, J. (2006). Hierarchical learning ensembles: Team building for undergraduate scientists and engineers. *Journal of College Science Teaching*, *35*(5), 21.

Dillman, D. A. (2006). Why choice of survey mode makes a difference. *Public Health Reports (1974-)*, *121*(1), 11-13.

Franklin, S., & Peat, M. (2005). Virtual versus real: An argument for maintaining diversity in the learning environment. *International Journal of Continuing Engineering Education and Lifelong Learning*, *15*(1-2), 67. doi:10.1504/IJCEELL.2005.006793

Harmon, H. L., & Wilborn, S. C. (2016). The math learning gap: Preparing STEM technicians for the new rural economy. *Rural Educator*, *37*(3), 26.

Hossain, M. M & Robinson, M. G. (2012). How to motivate US students to pursue STEM (Science, Technology, Engineering and Mathematics) careers. *US-China Education Review,* (4), 442-451.

HudsonAlpha Institute for Biotechnology. (2017). *Education*. Retrieved from https://hudsonalpha.org/education

Lewis, H. J. (2015). Mercedes-Benz U.S. International, Inc. (MBUSI). In *Encyclopedia of Alabama* online. Retrieved from http://www.encyclopediaofalabama.org/article/h-2065

Menshew, D. (2015). Using biotechnology, CSI, and zombies to promote science education in one of America's most challenging regions. *Journal of Commercial Biotechnology*, *21*(1), 69–77. doi:10.5912/jcb682

Mercedes-Benz, U. S. International. (2017). *Mercedes Manufacturing Excellence in Alabama*. Retrieved from www.mbusi.com/factory

Mosley, P., Ardito, G., & Scollins, L. (2016). Robotic cooperative learning promotes student STEM interest. *American Journal of Engineering Education*, *7*(2), 117–128.

National Academy of Sciences, National Academy of Engineering, and Institute of Medicine. (2011). *Rising above the gathering storm revisited: Rapidly approaching category 5*. Condensed version. Washington, DC: The National Academies Press. Retrieved from http://nap.edu/12999

National Aeronautics and Space Administration. (2016). *Todd May named Marshall Spaceflight Director.* Retrieved from https://www.nasa.gov/press-release/todd-may-named-marshall-space-flight-center-director

National Aeronautics and Space Administration. (n.d.). *Marshall Spaceflight Center*. Retrieved from https://www.nasa.gov/centers/marshall/overview.html

Netemeyer, R. G., Bearden, W. O., & Sharma, S. (2003). *Scaling procedures: Issues and applications*. Thousand Oaks, CA: Sage Publications. doi:10.4135/9781412985772

Simmons, K. (n.d.). *Kantis Simmons*. Retrieved from https://www.kantissimmons.com/p/about

Stuckey, M., Hofstein, A., Mamlok-Naaman, R., & Eilks, I. (2013). The meaning of 'relevance' in science education and its implications for the science curriculum. *Studies in Science Education*, *49*(1), 1–34. doi:10.1080/03057267.2013.802463

Thompson, J., & Soyibo, K. (2002). Effects of lecture, teacher demonstrations, discussion and practical work on 10th graders' attitudes to chemistry and understanding of electrolysis. *Research in Science & Technological Education*, *20*(1), 25–37. doi:10.1080/02635140220130902

U. S. Department of Health and Human Services. (n.d.). *Regina M. Benjamin (2009-2013)*. Retrieved from https://www.surgeongeneral.gov/about/previous/biobenjamin.html

U.S. Department of Education, Office of Innovation and Improvement. (2016). *STEM 2026: A vision for innovation in STEM education*. Washington, DC: Author. Retrieved from https://innovation.ed.gov/what-we-do/stem/

Ware, J., & Stein, S. (2013). From 'mentor' to 'role model': Scaling the involvement of STEM professionals through role model videos. *Journal of Educational Multimedia and Hypermedia*, *22*(2), 209–223.

KEY TERMS AND DEFINITIONS

Alabama Nature Center: Provides nature-based field trip opportunities where students can get hands-on experience in biological, chemical, and environmental sciences.

Career Exploration: Opportunities to examine STEM careers in depth including educational requirements, knowledge and skills needed, internships, etc.

Competitive STEM Events: Competitions which engage students in hands-on activities in science, technology, engineering, and mathematics.

Hudson Alpha Institute for Biotechnology: A collaborative biotechnology facility with a focus on genomic discoveries and promoting genomic literacy among the public.

Invention Convention: Supports student creativity and problem-solving. Students design, build and test inventions providing a real-world STEM experience.

Marshall Space Flight Center: A NASA facility responsible for the design and production of engines, instruments, payloads, and other products for spaceflight.

Mercedes-Benz U.S. International Incorporated: Located in Tuscaloosa, AL producing Mercedes-Benz vehicles: GLE SUV, GLS SUV, the C-Class and the GLE Coupe.

Professional Exploration Trips: Trips to STEM industry sites where students are allowed to interact with STEM professionals in their work environment.

Redstone Arsenal: A military installation that supports space exploration and U.S. missile defense.

STEM Challenge: A competitive event which involves students using skills in science, technology, engineering, and mathematics.

STEM Self-Efficacy: A student's belief in his/her knowledge and skills related to success in science, technology, engineering, and/or mathematics.

STEM Speakers: People who have distinguished themselves in an area of science, technology, engineering, and/or mathematics.

Team Building: Activities which foster cooperation and communication among individuals who work as a team.

U.S. Space and Rocket Center: A museum which showcases artifacts from the U.S. space program, simulation equipment, and hands-on labs for students.

Chapter 5
Opening the Doors to STEM Employment Opportunities for Rural Youth:
Guided Career Exploration

ABSTRACT

Guiding rural students on the path towards a STEM career requires a tripartite approach that addresses career decidedness, career thoughts, and vocational maturity. The authors provide an overview of a guided career exploration model employed by Project Engage. Through guided career exploration, rural students are introduced to the multitude of career possibilities in STEM as well as the knowledge and skills needed for those careers. Additionally, this chapter reviews the outreach component of Project Engage. Outreach to students from local high schools serves both as a recruitment tool and as a means to motivate rural students to pursue STEM as a college major. Finally, the authors present data from surveys regarding the effectiveness of the outreach activities on engaging high school students in STEM.

INTRODUCTION

Strategic mentoring as described in Chapter III increases the retention of undergraduates in STEM majors, yet, further guidance or career exploration is essential for directing college graduates into STEM careers. While career

DOI: 10.4018/978-1-5225-6341-9.ch005

planning and self-efficacy, as noted by Turner and Lapan (2002), consistently predicated career interest, Gnilka and Novakovic (2017) determined career search self-efficacy mediated women and non-white's perceived barriers to STEM careers. High schools serving low income populations often lack the resources for providing career counseling services (Schneider, Broda, Judy, & Burkander, 2013). Furthermore, rural students often lack realistic visions of what STEM careers entail as their experience with STEM professionals is limited. Project Engage uses a tripartite approach to career exploration to promote career decidedness, career thoughts, and vocational maturity. These three parts of the guided career exploration component of Project Engage support the development of STEM career self-efficacy and therefore increase the likelihood of students pursuing STEM careers in the future. The primary purpose of this chapter is to present the guided career exploration model employed by Project Engage.

In addition to supporting college-aged students, Project Engage works with high schools through an outreach component to introduce high school students to STEM majors and careers. A supply of students is needed to increase enrollment in STEM majors, thereby ultimately increasing the flow of STEM professionals into the workforce. Outreach to high schools is needed to ignite student interest in STEM during the secondary education years in order to increase enrollment in postsecondary STEM programs. The second purpose of this chapter is to describe how Project Engage capitalizes on popular television programming to create competitive events that garner secondary students' interest and motivate them to pursue STEM majors.

CAREER EXPLORATION: A TRIPARTITE APPROACH

As a capacity building project, Project Engage was intended to strengthen the STEM workforce pipeline by retaining students in their STEM programs and helping them make informed decisions to pursue STEM careers after graduation. Due to the isolation and lack of STEM industry in rural areas, students rarely interact with STEM professionals. This deficit results in a lack of vision regarding what a STEM career means. Career decision making can be stressful, especially for young people in emerging adulthood. According to Arnett (2007), emerging adulthood is a time of decision making and confusion for young people in their late teens and early twenties. In their emerging adulthood and facing the gaps of lacking STEM experience and role models, career decision making is stressful for STEM majors. Turner and Lapan (2002) found that career exploration/planning along with career self-efficacy are associated with student career interests. Along those lines, Project Engage provided a Guided Career Exploration program, illustrated in Figure 1, to STEM majors in order for them to become confident and skillful in career problem solving and career decision making.

The Guided Career Exploration program adopted a tripartite approach addressing STEM majors' career decidedness, career thoughts, and vocational maturity. Career decidedness is the degree to which individuals feel decided about their career choice (Lounsbury, Tatum, Chambers, Owens, & Gibson, 1999). Career decidedness can be viewed as a continuous variable on a scale from undecided to decided (Smith, 2011). Career thoughts include one's feelings, assumptions, attitudes, beliefs, behaviors, expectations, and plans or strategies related to career problem solving and career decision making (Peterson, Sampson, & Reardon, 1991; Sampson, Reardon, Peterson, & Lenz; 2004). Career thoughts are an important construct to address because negative career thoughts can hinder effective career problem solving and decision making (Paivandy, Bullock, Reardon, & Kelly, 2008; Sampson et al., 2004). Negative career thoughts are dysfunctional cognitions such as "I've messed up the best opportunity of my life" and "I'm never going to get another job that good," and such cognitions mediate and change individuals' career behaviors causing procrastination, anxiety, dependency, and/or premature foreclosure and consequently limiting the effectiveness of career problem solving and decision making (Chason, Bullock-Yowell, Sampson, Lenz, & Reardon, 2013). Vocational maturity is the ability to make informed and age-appropriate career decisions that fit one's abilities, occupational expectations, interests, and preferences and to successfully manage career development tasks. It has been seen as the collection of behaviors needed to identify, plan, and execute career goals (Prescod, 2014). It has been reported that college students with higher career maturity better realized their potential and demonstrated a higher degree of social adjustment than those with lower career maturity (Savickas, Briddick, & Watkins, 2002).

The tripartite approach of Guided Career Exploration utilized by Project Engage (see Figure 1) was aimed to: 1) enhance STEM students' career decidedness by providing STEM career information, having students study their own interests, and exploring STEM career opportunities; 2) cultivate positive STEM career thoughts by increasing students' self-efficacy about STEM careers and by identifying, challenging, and altering students' negative

STEM career thoughts; and 3) enhance students' vocational maturity by involving students in STEM career planning and development activities to improve their awareness of what is required to make an appropriate career choice and their ability to make career decisions that are both realistic and consistent over time.

GUIDED CAREER EXPLORATION ACTIVITIES

Youth from rural areas usually lack an understanding of what careers are possible as well as what educational and experience requirements are needed to achieve career goals. Project Engage freshmen STEM students are first introduced to STEM careers as part of their STEM 101 freshmen seminar class through professional field trips and invited speakers. Both activities provide the freshmen STEM students opportunities to investigate STEM industry and interact with STEM professionals fostering career decidedness. Professional exploration field trips to the Alabama Nature Center introduced students to careers as biologists, environmental engineers, and zoologists. The STEM professionals at Marshall Space Flight Center and HudsonAlpha Biotechnology guided students through the many facilities while providing information regarding educational requirements, internship opportunities

Figure 1. Tripartite approach of guided career exploration

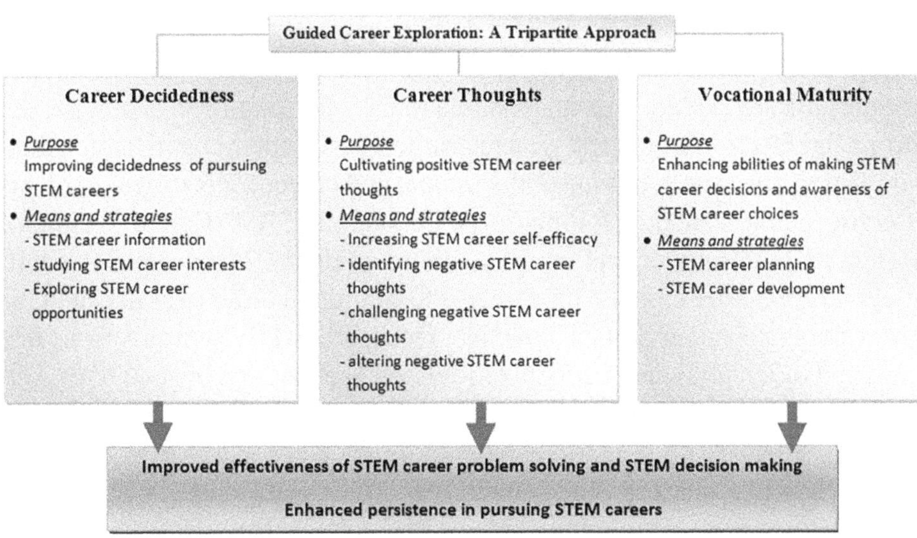

and more. The trips to the STEM professional sites introduced the students to endless possibilities for STEM careers.

In addition to field trips to STEM industry sites, the Project Team recruited STEM guest speakers to campus. Guest speakers have varied in their expertise including a former U.S. Surgeon General and current physician from rural Alabama, NASA project director, physicist, chemist, forensics expert, founder of the Minority Aviation Education Association and Interactive Science Program, and a STEM career motivational speaker. Together, these individuals have provided Project Engage participants with a holistic picture of the many possibilities for careers in STEM.

Career Exploration Course

The University also offers a freshmen seminar course (102- Career Explorations) that evaluates student interest and aptitude for different careers. The course additionally provides instruction on resume' writing, interview skills, graduate and other professional school requirements, and internship opportunities. The Project Team saw a need to expand the existing 102, general career explorations course, to provide guided STEM career exploration for STEM majors. This effort served a two-fold purpose: 1) allowed Project Engage freshmen to remain a cohesive group during the second semester of college, and 2) provided a more formal exploration of STEM careers. Project Engage expanded the concept of career exploration by adding a specific STEM focus to the course (102-STEM Career Explorations).

During the 102- STEM Career Explorations course, students are required to complete the Myers-Briggs Type Indicator® (MBTI®), an instrument for determining personality profile. Knowing their personality profile guides students toward careers which are suited to their personalities. Likewise, the Strong Interest Inventory® test is an instrument that students use to match their personal interests with possible educational and career trajectories. Students also use online career planning software such as SIGI[3] Career Guidance System and O*Net Occupational System to guide their career paths. SIGI[3] integrates self-assessment with career information in order to assist students in developing a career plan. The O*Net Occupational System functions in a similar manner, allowing students to research occupations in general as well as research careers based on their skill sets. Collectively these assessments provide students with holistic insights into the career that is best suited for their personality, interests, and career goals. Wang and Degol (2013) caution us not to force or coerce students into career paths that are not suited to their interests or abilities, but rather provide the information they need to make informed decisions.

STEM guest speakers, career fairs, as well as graduate and professional school recruitment events serve to further indoctrinate freshmen STEM majors to the opportunities available. Students who successfully complete the STEM career exploration course are able to do the following.

- Describe and use a model for making decisions and transitions that can be applied again and again throughout life.
- Describe personal characteristics identified using the STEM Careers Inventory (based on Holland Career Development Theory), Myers Briggs Type Indicator (MBTI) or Keirsey Temperament Sorter (Personality Inventory), Sigi3 computerized career guidance program, and various classroom activities.
- Identify a group of occupations that meet the criteria suggested by these characteristics and goals.
- Identify and select the training needed to enter one or more of the favored occupations.
- Present skills and documents that will assist in the job-search process.

Career Panel Discussion

The Career Panel Discussion was an event conducted as part of the career counseling provided to the Project Engage students. Project Engage invited a panel of STEM experts representing the STEM disciplines to campus. This panel met with the freshmen Engage students in a formal panel setting. The panel spoke briefly, introducing themselves and their occupations. The introductions were followed by a question and answer period for the students. This opportunity allowed students to interact with STEM professionals and gain first-hand knowledge of the panel's careers. The Career Panel Discussion data were collected via a brief survey following the event. In addition, Project Engage extended its guided career exploration to high school students through STEM Challenge activities. Data were collected from high schools about their experience in the STEM Challenge activities though a survey instrument employing a Likert-type scale.

Undergraduate Research

Nurturing positive "career thoughts" is the second part of the tripartite approach for career exploration (See Figure 5.1). Previous research (Carpi, Ronan, Falconer, & Lents, 2017) revealed that "...graduates participating in mentored research are increasingly focused on professional and academic STEM career tracks involving post-graduate study" (p.169). Undergraduate minority students in STEM programs gained self-efficacy by building skills as researchers. Participation in research by undergraduate students, in particular minority students, has benefits such as the mentor-student relationship and explicit career guidance and skill building. Project Engage has supported several STEM student researchers. Many participate in the institution's Undergraduate Research Symposium, while others travel to other institutions during the summer to participate in new research opportunities. Undergraduate research allows STEM majors to experience working as a STEM professional, but in a low pressure, non-threatening manner allowing for STEM self-efficacy to develop.

Sisters of the Academy "Priming the Pipeline"

One of the priorities of Project Engage was recruiting and retaining minority females into STEM. To that end, Project Engage has supported female African American STEM students attending Sisters of the Academy's Priming the Pipeline summer workshop. Sisters of the Academy intends to establish a network of Black women in higher education. This female organization offers a program called "Priming the Pipeline." According to the website of Sisters of the Academy (2015):

Priming the Pipeline *is a program focused on mentoring, leadership, and community engagement. Scholars of Sisters of the Academy mentor undergraduate and graduate women of color from various universities across the nation:*

- *Learn about themselves, particularly their educational and career goals;*
- *Identify their strengths and passions;*
- *Learn to work in collaboration with others with in an untraditional, necessary, And nurturing environment; and*

- *Through mentoring be steered in a positive direction to achieve positive future goals.*

Priming the Pipeline provides an opportunity for mentorship and encourages minority female STEM majors to build networking relationships.

Tripartite Approach Summary

Career decidedness (i.e., students' ability to make informed decisions regarding their future careers), and career thoughts (i.e., the beliefs, self-efficacy and expectations) are the first two parts of a tripartite approach to Guided Career Exploration. Several strategies can be employed in order to foster the development of career decidedness and positive career thoughts: providing wide-ranging STEM career information (STEM career exploration course); encouraging career exploration of STEM professions (field trips and guest speakers); and perhaps most importantly, examining personal interests and personality types (MBTI, Strong Interest Inventory, SIGI^3 Career Guidance System and O*Net Occupational System). Collectively these activities create a comprehensive approach to developing strong STEM career self-efficacy and to guiding students to STEM careers.

The third component of the tripartite approach to career exploration is "vocational maturity." The purpose of vocational maturity is to increase awareness of career opportunities and to enhance abilities to make informed career decisions. Working with the graduate career counselor, participants in Project Engage develop an e-portfolio showcasing their knowledge, skills and achievements in STEM. This final step in preparation for graduation and entrance into either graduate studies or a STEM career requires Project Engage participants to synthesize all that they have learned regarding careers, career choice, and their knowledge, skills and abilities in order to make a mature vocational decision regarding the next step on the pathway to a STEM career. Through implementation of Engage's tripartite approach to career exploration, students are capable of making informed decisions, reducing the chance of "leaking" from the pipeline to STEM career.

OUTREACH TO HIGH SCHOOL STUDENTS

Career exploration is essential to introducing students to the possibilities of careers in STEM, and therefore, should begin prior to college. Recruiting academically strong, highly motivated students into STEM fields is a priority if the U. S. is to grow its STEM workforce. By the time students reach the secondary level, interest in STEM typically declines. Focused efforts are needed to foster STEM interest and self-efficacy at the secondary level. According to McDonald (2016) "The implementation of inquiry-based practices that value active learning, immersion in authentic settings, engagement in reasoning and problem-solving, and the development of creativity, have been shown to facilitate effective student learning" (p. 551). Others have demonstrated that K-12 students are interested in authentic science experiences (Blankenburg, Höffler, & Parchmann, 2016) and hands-on STEM learning activities that positively impact student interest in STEM (Holstermann, Grube, & Bögeholz, 2009). One of the purposes of Project Engage is to recruit students, particularly minority students, into STEM undergraduate majors. To that end, Project Engage sponsors outreach activities for local rural high school students that include authentic STEM learning experiences. Furthermore, the authentic learning experiences promote STEM career decidedness and career thoughts in secondary students about future career choices in STEM.

High School Student Outreach: Invention Convention

Invention Convention is a national STEM competition that challenges students to use their creativity as well as their engineering design skills to "invent" a useful item and market it. Project Engage used the Invention Convention theme as a basis for its "Convention" STEM competitive event and as an outreach activity to high school students. Over the span of the grant, beginning in 2012, several versions of the Convention have tested student critical thinking, problem solving and creative abilities to "design" an item or items including a "green" creation from recycled materials, an egg drop carrier for a Shark Tank competition, boomerangs, and a multitude of other items. Project Engage's versions of the Invention Convention sparked high school students' interest in STEM, while building STEM self-efficacy as students prove to themselves they are up to the task.

Green Convention

During year one of the funding period, Project Engage sponsored an Invention Convention where college STEM students participated in robotics training and competition. Although the Engage Project Team saw the innate benefits of using robotics to further computational thinking, problem solving skills, programming and cooperative learning, there was not adequate time during year one to prepare the high school students for a robotics competition; therefore, an alternate STEM activity was chosen for their Invention Convention. Motivating students to pursue a STEM major in college by sparking an interest in STEM was still the priority. Since the overall theme for the competition was "Inventions," the Engage Project Team worked closely with the high school science and mathematics teachers to devise a "Green" Invention Challenge. The high school teams were tasked with taking an everyday object and recycling or repurposing it into a new product. The challenge required the high school students to use the skills they would need in a STEM related career: collaboration, creativity, design skills, presentation skills, and more. At the Invention Convention, the students presented their new products and described the process for creating it. The "Green" inventions were judged on creativity, design, and usefulness. Getting students excited about STEM early stands to impact future student decisions to pursue STEM majors in college.

Shark Tank Competition

By outsourcing the robotics component of the Invention Convention for the STEM undergraduates during year two, time for the Invention Convention on campus was solely dedicated to working with the local high school students. As a result, the college students, mentors and mentees, assumed leadership roles during the high school competition.

For year two, the Engage Project Team developed a "Shark Tank" theme for the Invention Convention, capitalizing on the popularity of the *Shark Tank* television show. On the *Shark Tank* television program, inventors appear before a panel of judges, "the sharks," where they present their inventions. The inventors must make a "sales pitch" for the invention and answer questions from "the sharks." For Project Engage's "Shark Tank," the sharks were mentors from Project Engage. The mentees were also given specific tasks to complete during the competition, such as creating STEM challenge videos for distribution to the high school in order to cultivate their critical thinking and problem solving skills. The goal of the event was to provide the Engage freshmen and mentors with leadership experience while simultaneously teaching the local high school students essential knowledge and skills needed to be successful in STEM.

The purpose of the Invention Convention was to reach out to local schools and provide high school students interaction with individuals outside the local schools. Distance between rural schools and any nearby towns, activities, businesses, etc. can often result in feelings of isolation. Interactions between the Project Engage participants and the local high school students were hypothesized to overcome some of the feelings of isolation and to impact student interest in STEM degrees. The Project Engage Team wanted to first get the high school students thinking mathematically and scientifically; however, the distance to the various schools prohibited the Engage participants from traveling to the high schools on a weekly basis. This led to the creation of a series of video challenges teams (e. g. Drops on a Penny, Microbiology, and Math Probability) which were posted on YouTube and delivered to the teachers via email over the course of four weeks. The purpose of the video challenges was to develop critical thinking and problem solving skills in the high school students. Mentors and their freshmen mentees created a series of videos which explained a scientific or mathematical concept and challenged the high school students to develop their own videos which demonstrated the same concept but using different methods, materials, or scenarios. The students recorded their videos and submitted them to the Engage Team for pre-competition judging. The more complex and creative the video demonstration, the higher the score. The highest scoring video of each high school team was then submitted to the "sharks," a group of Engage mentors for additional viewing. The "sharks" used the videos to determine what questions they planned to ask each team. The goal of the questions was to determine if the high school team members understood the concepts presented in their videos. Special awards were given based on the results of the "sharks" questioning of the high school student teams' videos. By using technology, Project Engage was able to overcome the distance obstacle.

The "Shark Tank" video competition initiated the event, allowing the high school students to become acclimated to the campus and more at ease interacting with the college students. After the video portion, the Invention Convention then entered the second portion, engineering an egg drop container that would keep an egg safe during a drop from a high distance. The engineering design process encompassed all areas of STEM plus additional simulated experience working as part of a research and development team. High school students were shown to their "Research and Development" lab and given a set amount of "money" to purchase all supplies that would be needed. A key aspect of the competition was the brainstorming and design process. Once the high school students determined the most effective "egg carrier" design, they had to do mathematical calculations to determine if the "materials" needed for their design were affordable. If the designs were over budget, they were required to make modifications, thus simulating a real-world research and development process. Once the design was within budget, students had to obtain their building materials from the supply store. The graduate career counselor for Project Engage served as the storekeeper while Engage freshmen assisted the high school students in finding the materials they planned to use. The next step was the follow through with the building of the "egg carrier" based on their design. The entire process took the students through an experience simulating the real world of a STEM researcher. This challenge required students to work collaboratively in their groups, design a container to protect the egg while simultaneously considering their limited resources, and finally to construct the protective container. The students were then required to test the effectiveness of their design and construction, completing the research and development process just as a STEM researcher would do. After scoring the egg drop containers, the high school students were questioned by the "sharks" about their team's design. The results of the questioning required the students to reflect on their designs and justify or explain their design and construction, another competence that real world STEM professionals must possess. The Shark Tank- themed Invention Convention gave the high school students experience working and competing as an engineering team and the STEM college students an opportunity to gain leadership experience.

Boomerang Convention

The Project Engage Team continued to improve on the Invention Convention competition in year three, inviting a world champion boomerang thrower and physicist to lead the competition. Math and science teachers from local participating schools were provided with an information packet on the principles behind boomerang flight in order to prepare their students in advance of the competition. Teachers were asked to use the boomerang lesson plans prior to attending the Invention Convention. Boomerangs sound like an unlikely theme for a STEM competitive event. However, boomerangs are deceptively scientific encompassing four physics principles including inertia, center of gravity, gyroscopic principles, and Bernoulli's Principles of air resistance, lift, pressure and relative velocity. Boomerangs are also especially engaging to young high school students due to the hands-on active learning involved. To begin the competition, the event leader provided an overview of the physics related to boomerang flight.

- **Inertia:** The law of inertia states that an object at rest will stay at rest unless acted upon by an unbalanced force.
- **Center of Gravity:** The center of gravity is the balance point of an object. Forces acting on the center of gravity create a torque.
- **Gyroscopic Principle:** The gyroscopic principle states that an applied force on a gyroscope will act a quarter of a rotation later.
- **Bernoulli's Principle:** When a fluid flows past an object perpendicularly it exerts a force on it called lift.

Following the educational period where the students from local high schools received boomerang physics from a physicist who happened to also be world champion boomerang thrower, students were challenged to design and build a boomerang which could outfly the boomerangs of their peers. Activities such as the boomerang competition are both intriguing and educational. Students see STEM in action in a nonthreatening manner which serves to lessen fear of subjects where students may not be confident. Engage's outreach to local schools from rural areas serves to recruit students into STEM programs.

Building a Better Invention

During year four of Project Engage a former rocket scientist, author, and nationally renowned speaker was invited to lead the high school students' Invention Convention. Year four's leader, a successful African American male who was also a very highly qualified and experienced STEM professional (former NASA Langley research scientist) turned academic success motivational speaker, was a perfect choice for the event. The Invention Convention required students to use their STEM skills to complete four different events. The events required students to build four distinctly different objects: a balloon rocket, a tabletop hovercraft, a soap-powered motorboat and a film canister rocket. A general description of the objects follows.

- **Balloon Rocket:** The balloon rocket demonstrates the concept of thrust used to propel rockets. Students first observed a simple demonstration and then created their own balloon rocket using a balloon, straw, and string.

- **Tabletop Hovercraft:** The tabletop hovercraft consists of a balloon, a pop-top cap from a dish detergent bottle, and a compact disc (cd). The hovercraft demonstrates the concept of air flow.
- **Soap-Powered Motorboat:** The soap-powered motorboats were made of plastic trays or cardboard. Boats were placed in a tray of water with a drop of soap added to cause the boat to propel demonstrating the ability of soap to act as a surfactant (something that breaks surface tension of water).
- **Film Canister Rocket:** A film canister rocket was constructed using an empty film canister and an Alka-Seltzer tablet. The film canister rocket demonstrated both the concept of pressure created by gas in a contained area as well as the idea of thrust.

However, the experience did not end with the building. Engineers rarely design and build without testing and modifications to perfect their creations; thus, teams were also required to make modifications to the object, test the various modifications and record their results. For example, students were required to modify the shape of the balloon, length of the straw, type of string, and angle of string to determine the effects on thrust. Other examples of modifications made by the students included changing the balloon size, the type of gas in the balloon, and the size of the disc used in the hovercraft. Students were conducting experiments on the hovercraft design to determine which modifications made the craft hover best. Additionally, students modified the soap-powered boats by altering the material the boat was made of, the type of soap used (liquid or bar), and the temperature of the water to see what had the greatest effect on propulsion. Finally, in the fourth exercise, students made modifications to the film canister rocket and tested the effects on thrust. Several variables were tested including the effect of water temperature on speed of the rocket launch, controlling the flight path by adding fins or a nosecone to the canister, what amount of water in the canister will give the highest flight and quickest launch of the rocket. The teams gained both knowledge of scientific principles through building and testing the effects of the modifications to the objects as well as experience working as a project team member, conducting scientific experiments, and drawing conclusions (critical thinking). Year four's Invention Convention not only taught needed STEM skills and ways of working in STEM fields, the motivational speaker also inspired students to concentrate on their academic pursuits. The purpose of the Invention Convention extended beyond outreach to high schools and recruitment of STEM majors to providing motivation through hands-on, inquiry science.

High School Student Outreach: STEM Challenge

STEM Challenge: Forensics

Due to the success of the Forensics Challenge in year one, the Engage Project Team decided to repeat the theme, but with a new crime scene scenario for year two. For year two, only the high school teams competed in the event; however, the Engage college freshmen and mentors were given leadership roles as team leaders, actors in the crime scene scenarios, and workers at the individual crime scenes. The primary activity of the STEM Challenge was the crime scene scenario involving a love triangle gone wrong. In the scenario, the victim (a male student dating two different female students) passed out while walking to class drinking an energy drink. To add a touch of authenticity to the event, campus police were surrounding the body as if they just arrived on the scene. Student response to the police on scene was favorable. It gave the crime scene the appearance of a real world experience. On the day of the competition, the college freshmen teams and the high school student teams were introduced to the "crime" and tasked with visiting multiple sites to collect evidence. Evidence provided to the students included: 1) a fingerprint from the energy drink can; 2) chemical detecting poison in the energy drink; 3) pictures of bruises on the male victim and one of the female girlfriends; 3) suspect interrogation; and 4) a letter read to one of the suspects with a secret code in letter between the girlfriend and the victim discussing their relationship. After all of the "evidence" had been collected, teams were challenged to solve the "crime" by analyzing the evidence. Teams had to have knowledge of fingerprint analysis techniques in order to match fingerprints collected to the various suspects. In addition to the fingerprint analysis, participants had to have a working knowledge of chemical analysis techniques so as to correctly identify the "poison" in the energy drink. For this year's scenario, the Team added an interview portion to the challenge, so that student participants would have to use their listening skills for added insight into the case. The interrogations also served the secondary purpose of requiring students to base their conclusions on hard evidence. Finally, the participants had to create a presentation of their case using the photos taken at the various evidence sites to support their theory of the crime. The high school students (and the Engage college participants) get very excited about the crime scene investigations and hopefully about STEM as well.

STEM Challenge: S-T-E-M Round Robin

For year three, the team doubled its efforts to concentrate on outreach to rural schools by inviting a guest speaker, a prominent African American STEM specialist in chemistry and math as well as math and science education, to visit various local rural schools bringing his interactive science show to the students. Students from schools located in remote rural areas rarely meet scientists, especially African American scientists. The guest speaker inspired numerous students both as a role model to whom they could relate and by getting them out of the audience and interacting with his science demonstrations. The interactive science demonstrations served to inspire high school teams to attend the third annual STEM Challenge. For this year, rather than hosting a forensics competition, the Engage Project team took advantage of the speaker's expertise and allowed him to develop the STEM Challenge activities including a cow heart dissection (science), robotics programming competition (technology), a dominos design competition (engineering), and a math problem involving complex calculations (mathematics). Student teams rotated through the various events accumulating teamwork and STEM skills.

The science event, a cow heart dissection allowed the high school students to have a hands on science lab experience, one that underfunded rural schools cannot always provide. Several studies have claimed the benefits of hands-on science to promote student academic achievement (Ates & Eryılmaz, 2011; Bigler & Hanegan, 2011; Caglak, 2016). Project Engage offers high school students the opportunity for hands-on laboratory experience with the goal of increasing student academic achievement as well as interest in STEM.

As today's need for employees skilled in computing increases, programming skills must be taught early. The technology component of the competition involved the student teams working to program robots to complete certain tasks. This STEM Challenge in year three was Project Engage's first effort in providing robotics education to high school students. The teams were first given a brief introduction to how programming works to control the robots. For this event, the robots were controlled by programmable calculators. After the introduction, the teams were then given time to complete the programming of the robots to move through a maze with turns and, finally, compete against each other. The goal of this activity was two-fold: build teamwork skills and increase student interest in careers in computer science and programming.

Creativity and design skills are also prime skills needed to be successful in STEM. Engineering design has the potential to promote a scientific way of approaching and solving problems. Siew, Goh and Sulaiman (2016) studied the effect of implementing an engineering design program on rural secondary students. The results indicated the engineering design program supported the development of critical thinking, problem solving, and creativity while increasing the rural students' awareness of their own potential in STEM. Project Engage serves a similar community of students, rural and secondary level, and hoped to duplicate the results of Siew et al. (2016). The domino challenge of the competition involved teams creating an original design and building a domino structure that would completely fall with one push. Students had to be knowledgeable about action-reaction forces, angles, energy, and other physics related science principles. Student designs were judged on creativity and effectiveness in achieving the goal.

The final STEM Challenge was a mathematical problem solving activity called the Power of Ten. The activity required students to think "outside the box," extrapolating from the size of an area on a map to the size of the universe expressed in powers of ten. Computational thinking (CT) as defined by Yadav, Stephenson and Hai (2017) "… is a set of problem-solving thought processes derived from computer science but applicable in any domain" (p.56). High school students involved in the mathematics challenge were required to use CT to express the size of the universe in powers of ten. Using such CT skills is challenging for most students, especially those from rural areas who most likely do not receive training in using CT. The results of this challenge supported the need for continued CT education.

STEM Challenge: Return to Forensics

Results from the year three surveys as well as informal discussions indicated that the high school students related to the forensics theme, preferring it over the individual STEM activities. The Project team continually revises the competitive events to improve their appeal to the high school students and their effectiveness in building STEM knowledge and skills. Forensic investigations provide opportunities for the application of all types of scientific and mathematical knowledge and skills. For year four, the forensics challenge was created by an actual forensics scientist from Alabama's Department of Forensics providing an even more authentic learning experience. The forensics scenario required students to determine if a drug overdose of a fictional victim was an accident, attempted suicide or attempted homicide. The high school teams were required to 1) determine what drugs were found in the victim using a drug analysis kit; 2) determine how much of the drug is required to overdose by using mathematical calculations, 3) analyze hair samples found on the scene using a hair analysis kit, 4) and analyze the case file of the unresponsive victim, including the background information, medical history, family history, academics, emails, and text messages. From the physical evidence and witness interviews, the high school student teams were challenged to first, determine if the overdose was accidental, attempted suicide or attempted homicide. Secondly, if the team ruled the overdose a homicide, the suspected perpetrator had to be identified along with a motive. Finally, the teams had to present their findings to the panel of judges. Students were required to convince the panel that their evidence supported the sequence of events and motive (if applicable). Every argument and/or statement presented had to be supported with evidence. Findings from Project Engage's Forensics STEM Challenges and research by Marle et al. (2014) found that crime scene based scenarios allowed for the introduction of advanced level scientific concepts and practices in a way relevant to students of today.

RESEARCH FINDINGS FROM GUIDED CAREER EXPLORATION ACTIVITIES

Career Panel Discussion Survey

The Career Panel Discussion survey data was available from the 2013 Cohort (n = 18). The instrument used to collect the survey data consisted of demographics questions about race, ethnicity, and STEM majors, and survey items regarding the effects of the Career Panel Discussion on the undergraduate students' interest in STEM fields, knowledge of STEM fields, persistence as STEM majors, and intention to work as STEM professionals after graduation. The demographic information of the survey respondents is presented in Table 1. The survey items were positively worded to avoid confusion for respondents (Netemeyer, Bearden, & Sharma, 2003) and cognitive burdens for information processing (Dillman, 2006). The survey items adopted a 4 point Likert-like scale (i.e., strongly disagree, disagree, agree, and strongly agree). In data

Table 1. Demographic information of the Career Panel Discussion survey participants

Gender	Not available
Ethnicity	Hispanic/Latino: 0
	Not Hispanic/Latino: 18
Race	African American: 12
	White: 5
	American Indian or Alaskan Native: 1
Major	Science: 13
	Technology: 4
	Engineering: 1
	Mathematics: 0

analysis, the responses to the survey items were coded with "1 = strongly disagree", "2 = disagree", "3 = agree", and "4 = strongly agree".

Interest in STEM Fields

The mean to the survey item "Overall, the STEM Professional Career Panel Discussion increased my interest in STEM fields through career counseling" was 3.33 indicating a very positive rating of the effects of the STEM Professional Career Panel Discussion on increasing the students' interest in STEM fields. The frequency distribution of the non-white and white students' responses is presented in Table 2.

Due to the small sample size of the survey participants and the disproportionate number of the non-white and white students, the "within

Table 2. Career Panel Discussion: Interest in STEM

Race	Reported Statistics	Disagree	Agree	Strongly Agree
White	Count (n)	1	3	1
	% Within Race	20.0%	60.0%	20.0%
	% Within Responses	100.0%	30.0%	14.3%
	% of Total	5.6%	16.7%	5.6%
Non-White	Count (n)	0	7	6
	% Within Race	0.0%	53.8%	46.2%
	% Within Responses	0.0%	70.0%	85.7%
	% of Total	0.0%	38.9%	33.3%

race" and "within responses" percentages are not able to speak much about the patterns of the two groups' responses. The mean of the non-white students' response was 3.46 and that of the white students was 3.00. A Mann-Whitney U test was conducted comparing the non-white and white students' responses and the result did not indicate a significant difference.

Knowledge in STEM Fields

The mean to the survey item "Overall, the STEM Professionals Career Panel Discussion increased my knowledge of STEM fields" was 3.33 indicating a very positive rating of the effects of the STEM Professional Career Panel Discussion on increasing the students' knowledge in STEM fields. The frequency distribution of the non-white and white students' responses is presented in Table 3.

Table 3 shows that most of the students (17 out 18) responded "agree" or "strongly agree" to the survey item. The mean of the non-white students' response was 3.38 and that of the white students was 3.20. A Mann-Whitney U test was conducted comparing the non-white and white students' responses and the rest result did not indicate a significant difference.

Persistence as STEM Majors

The mean to the survey item "As a result of participating in the STEM Professionals Career Panel Discussion, I am inclined to continue to pursue a STEM discipline as a major in college" was 3.44 indicating a very positive rating of the effects of the STEM Professional Career Panel Discussion

Table 3. Career Panel Discussion: Knowledge in STEM fields

Race	Reported Statistics	Disagree	Agree	Strongly Agree
White	Count (n)	1	1	3
	% Within Race	20.0%	20.0%	60.0%
	% Within Responses	100.0%	11.1%	37.5%
	% of Total	5.6%	5.6%	16.7%
Non-White	Count (n)	0	8	5
	% Within Race	0.0%	88.9%	38.5%
	% Within Responses	0.0%	70.0%	62.5%
	% of Total	0.0%	44.4%	27.8%

Table 4. Career Panel Discussion: Persistence as a STEM major

Race	Reported Statistics	Disagree	Agree	Strongly Agree
White	Count (n)	1	1	3
	% Within Race	20.0%	20.0%	60.0%
	% Within Responses	100.0%	14.3%	30.0%
	% of Total	5.6%	5.6%	16.7%
Non-White	Count (n)	0	6	7
	% Within Race	0.0%	46.2%	53.8%
	% Within Responses	0.0%	85.7%	70.0%
	% of Total	0.0%	33.3%	38.9%

on enhancing the students' persistence as STEM majors. The frequency distribution of the non-white and white students' responses is presented in Table 4.

Table 4 shows that most of the students (17 out 18) responded "agree" or "strongly agree" to the survey item. The mean of the non-white students' response was 3.54 and that of the white students' response was 3.20. A Mann-Whitney U test for comparing the non-white and white students' responses did not indicate a significant difference.

Intention to Work as STEM Professionals After Graduation

The mean to the survey item "As a result of participating in the STEM Professionals Career Panel Discussion, I am inclined to continue to pursue a STEM professional career after I complete college" was 3.44 indicating a very positive rating of the effects of the STEM Professional Career Panel Discussion on strengthening the students' intentions to work as STEM professionals after graduation. The frequency distribution of the non-white and white students' responses is presented in Table 5.

Table 5 shows that most of the students (17 out 18) responded "agree" or "strongly agree" to the survey item regarding their plans for a STEM career after graduation. The mean of the non-white students' response was 3.54 and that of the white students was 3.20. A Mann-Whitney U test for comparing the non-white and white students' responses did not yield a significant difference.

Table 5. Career Panel Discussion: Intention to work as a STEM professional

Race	Reported Statistics	Disagree	Agree	Strongly Agree
White	Count (n)	1	1	3
	% Within Race	20.0%	20.0%	60.0%
	% Within Responses	100.0%	14.3%	30.0%
	% of Total	5.6%	5.6%	16.7%
Non-White	Count (n)	0	6	7
	% Within Race	0.0%	46.2%	53.8%
	% Within Responses	0.0%	85.7%	70.0%
	% of Total	0.0%	33.3%	38.9%

High School Invention Convention Surveys 2013, 2014, 2015

Project Engage sponsored an Invention Convention among high school students in 2013, 2014, and 2015. Survey data were collected from the participating high school students.

The instrument used to collect the survey data consisted of demographic questions about gender, ethnicity, and race, and survey items regarding the effects of the Invention Convention on the high school students' interest in STEM fields, knowledge of STEM fields, intentions to enroll as STEM majors, and intentions to work as STEM professionals after college. Demographic information of the high school students is shown in Table 6. The survey items were positively worded to avoid confusion for respondents (Netemeyer, Bearden, & Sharma, 2003) and cognitive burdens for information processing (Dillman, 2006). The survey items adopted a 4 point Likert-like scale (i.e., strongly disagree, disagree, agree, and strongly agree). In data analysis, the responses to the survey items were coded with "1 = strongly disagree", "2 = disagree", "3 = agree", and "4 = strongly agree".

2013 High School Shark Tank Competition

Interest in STEM Fields

The mean to the survey item "Overall, the Invention Convention competition strengthened my interest in STEM fields" was 3.69 indicating a very positive

Table 6. Demographic information of the high school student participants

	2013 (n = 16)	2014 (n = 16)
Gender	Female: 8	Female: 5
	Male: 8	Male: 11
Ethnicity	Hispanic/Latino: 0	Hispanic/Latino: 0
	Not Hispanic/Latino: 16	Not Hispanic/Latino: 16
Race	African American: 8	African American: 6
	White: 8	White: 10
	2015 (n = 33)	
Gender	Female: 7	
	Male: 26	
Ethnicity	Hispanic/Latino: 0	
	Not Hispanic/Latino: 33	
Race	African American: 31	
	Asian: 2	

rating of the effects of 2013 Invention Convention on increasing the high students' interest in STEM fields. The frequency distribution of the non-white and white students' responses is presented in Table 7.

The "within race" percentages in Table 7 indicate that the white high school students were more likely than their non-white counterparts to "strongly agree" with the statement. The mean of the non-white high school students' response was 3.63 and that of the white high school students was 3.75. A Mann-Whitney U test was conducted comparing the non-white and white students' responses and the test result did not indicate a significant difference.

Table 7. Shark Tank Convention: Interest in STEM

Race	Reported Statistics	Agree	Strongly Agree
White	Count (n)	2	6
	% Within Race	25.0%	75.0%
	% Within Responses	40.0%	54.5%
	% of Total	12.5%	37.5%
Non-White	Count (n)	3	5
	% Within Race	37.5%	62.5%
	% Within Responses	60.0%	45.5%
	% of Total	18.8%	31.3%

Knowledge in STEM Fields

The mean to the survey item "Overall the Invention Convention strengthened my knowledge of STEM fields" was 3.50 indicating a very positive rating of the effects of the 2013 Invention Convention on increasing the high school students' knowledge in STEM fields. The frequency distribution of the non-white and white students' responses is presented in Table 8.

The "within race" percentages in Table 8 indicate that the non-white high school students, compared with the white high school students, were more likely to "strongly agree" with the statement. The mean of the non-white high school students' response was 3.75 and that of the white high school students was 3.25. A Mann-Whitney U test was conducted comparing the non-white and white students' responses and the test result indicated a significant difference ($z = 1.943$, $p = .049$).

Intention to Enroll as STEM Majors

The mean to the survey item "As a result of participating in the Invention Convention, I am inclined to pursue a STEM discipline as a major in college" was 3.25 indicating a positive rating of the effects of the 2013 Invention Convention on enhancing the high school students' intentions to enroll as STEM majors. The frequency distribution of the non-white and white students' responses is presented in Table 9.

The "within race" percentages in Table 9 did not give a clear picture about the preference of non-white and white high school students' rating of the survey item. The mean of the non-white and white high school students'

Table 8. Shark Tank Convention: Knowledge of STEM fields

Race	Reported Statistics	Agree	Strongly Agree
White	Count (n)	6	2
	% Within Race	75.0%	25.0%
	% Within Responses	75.0%	25.0%
	% of Total	37.5.5%	12.5%
Non-White	Count (n)	2	6
	% Within Race	25.0%	75.0%
	% Within Responses	25.0%	75.0%
	% of Total	12.5%	37.5%

Table 9. Shark Tank Convention: Intention to enroll as a STEM major

Race	Reported Statistics	Disagree	Agree	Strongly Agree
White	Count (n)	1	5	2
	% Within Race	12.5%	62.5%	25.0%
	% Within Responses	100.0%	50.0%	40.0%
	% of Total	6.3%	31.3%	12.5%
Non-White	Count (n)	0	6	7
	% Within Race	0.0%	62.5%	37.5%
	% Within Responses	0.0%	50.0%	60.0%
	% of Total	0.0%	31.3%	18.8%

responses was 3.38 and 3.13 respectively. A Mann-Whitney U test was conducted comparing the non-white and white students' responses, and the test result did not indicate a significant difference.

Intention to Work as STEM Professionals After College

The mean to the survey item "As a result of participating in the Invention Convention, I am inclined to continue to pursue a STEM professional career after I complete college" was 3.13 indicating a positive rating of the effects of the 2013 Invention Convention on enhancing the high school students' intention to work as STEM professionals after college. The frequency distribution of the non-white and white students' responses is presented in Table 10.

The "within race" percentages regarding intention to work as a STEM professional did not give a clear picture about the preference of non-white

Table 10. Shark Tank Convention: Intention to work as a STEM professional

Race	Reported Statistics	Disagree	Agree	Strongly Agree
White	Count (n)	1	5	2
	% Within Race	12.5%	62.5%	25.0%
	% Within Responses	25.0%	83.3%	33.3%
	% of Total	6.3%	31.3%	12.5%
Non-White	Count (n)	3	1	4
	% Within Race	37.5%	12.5%	50.0%
	% Within Responses	75.0%	16.7%	66.7%
	% of Total	18.8%	6.3%	25.0%

and white high school students' rating of the survey item. The means of the non-white and white high school students' responses were the same, 3.13.

2014 High School Boomerang Convention

Interest in STEM Fields

The mean to the survey item "Overall, the Invention Convention competition strengthened my interest in STEM fields" was 3.38 indicating a positive rating of the effects of the 2014 Invention Convention on increasing the high school students' interest in STEM fields. The frequency distribution of the non-white and white students' responses is presented in Table 11.

The "within race" percentages in Table 11 indicate that the non-white high school students were more likely than their white counterparts to "strongly agree" with the statement. The mean of the non-white high school students' responses was 3.67 and that of the white high school students was 3.20. A Mann-Whitney U test was conducted comparing the non-white and white students' responses and the test result did not indicate a significant difference.

Knowledge in STEM Fields

The mean to the survey item "Overall the Invention Convention strengthened my knowledge of STEM fields" was 3.31 indicating a positive rating of the effects of the 2014 Invention Convention on increasing the high school students' knowledge in STEM fields. The frequency distribution of the non-white and white students' responses is presented in Table 12.

Table 11. Boomerang Convention: Interest in STEM fields

Race	Reported Statistics	Agree	Strongly Agree
White	Count (n)	8	2
	% Within Race	80.0%	20.0%
	% Within Responses	80.0%	33.3%
	% of Total	50.0%	12.5%
Non-White	Count (n)	2	4
	% Within Race	33.3%	66.7%
	% Within Responses	20.0%	66.7%
	% of Total	12.5%	25.0%

Table 12. Boomerang Convention: Knowledge in STEM fields

Race	Reported Statistics	Disagree	Agree	Strongly Agree
White	Count (n)	1	6	3
	% Within Race	10.0%	60.0%	30.0%
	% Within Responses	100.0%	66.7%	50.0%
	% of Total	6.3%	37.5%	18.8%
Non-White	Count (n)	0	3	3
	% Within Race	0.0%	50.0%	50.0%
	% Within Responses	0.0%	33.3%	50.0%
	% of Total	0.0%	18.8%	18.8%

The "within race" percentages in Table 12 indicate that the non-white high school students, compared with the white high school students, were more positive about the survey statement. The mean of the non-white high school students' responses was 3.50 and that of the white high school students was 3.20. A Mann-Whitney U test was conducted comparing the non-white and white students' responses, and the test result did not indicate a significant difference.

Intention to Enroll as STEM Majors

The mean to the survey item "As a result of participating in the Invention Convention, I am inclined to pursue a STEM discipline as a major in college" was 2.81 not indicating a positive rating of the effects of the 2014 Invention Convention on enhancing the high school students' intentions to enroll as STEM majors. The frequency distribution of the non-white and white students' responses is presented in Table 13.

The "within race" percentages in Table 13 indicate that the non-white high school students had a more positive opinion about the survey statement. The means of the non-white and white high school students' responses were 3.17 and 2.60 respectively. A Mann-Whitney U test was conducted comparing the non-white and white students' responses, and the test result did not indicate a significant difference.

Table 13. Boomerang Convention: Intention to enroll as a STEM major

Race	Reported Statistics	Strongly Disagree	Disagree	Agree	Strongly Agree
White	Count (n)	1	4	3	2
	% Within Race	10.0%	40.0%	30.0%	20.0%
	% Within Responses	100.0%	66.7%	75.0%	40.0%
	% of Total	6.3%	25.0%	18.8%	12.5%
Non-White	Count (n)	0	2	1	3
	% Within Race	0.0%	33.3%	16.7%	50.0%
	% Within Responses	0.0%	33.3%	25.0%	60.0%
	% of Total	0.0%	12.5%	6.3%	18.8%

Intention to Work as STEM Professionals After College

The mean to the survey item "As a result of participating in the Invention Convention, I am inclined to continue to pursue a STEM professional career after I complete college" was 2.88 not indicating a positive rating of the effects of the 2014 Invention Convention on enhancing the high school students' intention to work as STEM professionals after college. The frequency distribution of the non-white and white students' responses is presented in Table 14.

The "within race" percentages in Table 14 indicate that the non-white high school students had a more positive opinion about the survey statement. The means of the non-white and white high school students' responses were 3.33

Table 14. Boomerang Convention: Intention to work as a STEM professional

Race	Reported Statistics	Strongly Disagree	Disagree	Agree	Strongly Agree
White	Count (Nn	1	4	3	2
	% Within Race	10.0%	40.0%	30.0%	20.0%
	% Within Responses	50.0%	100%	75.0%	33.3%
	% of Total	6.3%	25.0%	18.8%	12.5%
Non-White	Count (n)	1	0	1	4
	% Within Race	16.7%	0.0%	16.7%	66.7%
	% Within Responses	50.0%	0.0%	25.0%	66.7%
	% of Total	6.3%	0.0%	6.3%	25.0%

and 2.60 respectively. A Mann-Whitney U test was conducted comparing the non-white and white students' responses, and the test result did not indicate a significant difference.

2015 High School Building a Better Invention

Interest in STEM Fields

The mean to the survey item "Overall, the Invention Convention competition strengthened my interest in STEM fields" was 3.64 indicating a very positive rating of the effects of the 2015 Invention Convention on increasing the high school students' interest in STEM fields. All high school students in the 2015 Invention Convention were non-white students. The frequency distribution of these non-white students' responses is presented in Table 15.

Knowledge in STEM Fields

The mean to the survey item "Overall the Invention Convention strengthened my knowledge of STEM fields" was 3.03 indicating a positive rating of the effects of the 2015 Invention Convention on increasing the high school students' knowledge in STEM fields. All high school students participating in the 2015 Invention Convention were non-white. The frequency distribution of these non-white students' responses is presented in the following Table 16.

Table 15. Building a better invention: Interest in STEM fields

	Frequency	Relative Frequency
Disagree	2	6%
Agree	8	24%
Strongly Agree	23	70%

Table 16. Building a better invention: Knowledge in STEM fields

	Frequency	Relative Frequency
Disagree	3	9%
Agree	26	79%
Strongly Agree	4	12%

Intention to Enroll as STEM Majors

The mean to the survey item "As a result of participating in the Invention Convention, I am inclined to pursue a STEM discipline as a major in college" was 3.09 indicating a positive rating of the effects of the 2015 Invention Convention on enhancing the high students' intentions to enroll as STEM majors. All high school students in 2015 Invention Convention were non-white students. The frequency distribution of these non-white students' responses is presented in Table 17.

Intention to Work as STEM Professionals After College

The mean to the survey item "As a result of participating in the Invention Convention, I am inclined to continue to pursue a STEM professional career after I complete college" was 3.06 indicating a positive rating of the effects of the 2015 Invention Convention on enhancing the high school students' intention to work as STEM professionals after college. All high school students participating in the 2015 Invention Convention were non-white students. The frequency distribution of these non-white students' responses is presented in Table 18.

Table 17. Building a better invention: Intention to enroll as a STEM major

	Frequency	Relative Frequency
Disagree	7	21%
Agree	16	49%
Strongly Agree	10	30%

Table 18. Building a better invention: Intention to work as a STEM professional

	Frequency	Relative Frequency
Disagree	10	30%
Agree	11	33%
Strongly Agree	12	37%

High School STEM Challenge Surveys: 2013, 2014, 2016

The STEM Forensic Challenge was conducted among high school students in 2013, 2014, and 2016. Survey data collected from these three groups of high school students were combined for analyzation. Demographic information of the high school students is shown in Table 19. The instrument used to collect the survey data consisted of demographics questions about gender, ethnicity, and race, and survey items regarding the effects of the STEM Forensic Challenge on the undergraduate students' interests in STEM fields, knowledge of STEM fields, intentions to enroll as STEM majors, and intentions to work as STEM professionals after college. The survey items were positively worded to avoid confusion for respondents (Netemeyer, Bearden, & Sharma, 2003) and cognitive burdens for information processing (Dillman, 2006). The survey items adopted a 4 point Likert-like scale (i.e., strongly disagree, disagree, agree, and strongly agree). In data analysis, the responses to the survey items were coded with "1 = strongly disagree", "2 = disagree", "3 = agree", and "4 = strongly agree".

Table 19. Demographics of 2013, 2014, and 2016 high school STEM Challenge survey participants

	2013 (n = 21)	**2015 (n = 23)**
Gender	Female: 7	Female: 5
	Male: 14	Male: 18
Ethnicity	Hispanic/Latino: 0	Hispanic/Latino: 0
	Not Hispanic/Latino: 21	Not Hispanic/Latino: 23
Race	African American: 21	African American: 24
	White: 0	White: 0
	2016 (n = 41)	
Gender	Female: 20	
	Male: 21	
Ethnicity	Hispanic/Latino: 0	
	Not Hispanic/Latino: 41	
Race	African American: 18	
	White: 21	
	Asian: 2	

Interest in STEM Fields

The mean to the survey item "The STEM Forensic Challenge increased my interest in STEM fields" was 2.89 not indicating a very positive rating of the effects of the STEM Forensic Challenge on increasing the high students' interest in STEM fields. The frequency distribution of the non-white and white students' responses is presented in Table 20.

The "within race" percentages in Table 20 indicate the white high school students were more likely than their non-white counterparts to disagree with the statement. The mean of the non-white high school students' responses was 2.95 and that of the white high school students was 2.71. A Mann-Whitney U test was conducted comparing the non-white and white students' responses, and the test result did not indicate a significant difference.

Knowledge in STEM Fields

The mean to the survey item "The STEM Forensic Challenge increased my interest in STEM fields" was 2.92 not indicating a very positive rating of the effects of the STEM Forensic Challenge on increasing the high school students' knowledge in STEM fields. The frequency distribution of the non-white and white students' responses is presented in Table 21.

The "within race" percentages in Table 21 indicate the white high school students, with 71.4% in the "disagree" category, were less positive in their opinion about the effects of the Forensic Challenge on increasing knowledge in STEM fields. The mean of the non-white high school students' responses was

Table 20. STEM Challenge: Interest in STEM fields

Race	Reported Statistics	Strongly Disagree	Disagree	Agree	Strongly Agree
White	Count (n)	0	11	5	5
	% Within Race	0.0%	52.4%	23.8%	23.8%
	% Within Responses	0.0%	40.7%	13.5%	25.0%
	% of Total	0.0%	12.9%	5.9%	5.9%
Non-White	Count (n)	1	16	32	15
	% Within Race	1.6%	25.0%	50.0%	23.4%
	% Within Responses	100.0%	59.3%	86.5%	75.0%
	% of Total	1.2%	18.8%	37.6%	17.6%

Table 21. STEM Challenge: Knowledge in STEM fields

Race	Reported Statistics	Disagree	Agree	Strongly Agree
White	Count (n)	15	1	6
	% Within Race	71.4%	4.8%	23.8%
	% Within Responses	51.7%	2.9%	22.7%
	% of Total	17.6%	1.2%	5.9%
Non-White	Count (n)	14	33	17
	% Within Race	21.9%	51.6%	26.6%
	% Within Responses	48.3%	97.1%	77.3%
	% of Total	16.5%	38.8%	20.0%

3.05 and that of the white high school students was 2.52. A Mann-Whitney U test was conducted comparing the non-white and white high school students' responses, and the test result indicated a significant difference ($z = 2.815$, $p = .005$).

Intention to Enroll as STEM Majors

The mean to the survey item "As a result of participating in the Forensic Challenge, I am inclined to pursue a STEM discipline as a major in college" was 2.65 not indicating a very positive rating of the effects of the STEM Forensic Challenge on enhancing the high students' intention to enroll as STEM majors. The frequency distribution of the non-white and white students' responses is presented in Table 22.

Table 22. STEM Challenge: Intention to enroll as a STEM major

Race	Reported Statistics	Strongly Disagree	Disagree	Agree	Strongly Agree
White	Count (n)	0	9	11	1
	% Within Race	0.0%	42.9%	52.4%	4.8%
	% Within Responses	0.0%	27.3%	29.7%	9.1%
	% of Total	0.0%	10.6%	12.9%	1.2%
Non-White	Count (n)	4	24	26	10
	% Within Race	6.3%	37.5%	40.6%	15.6%
	% Within Responses	100.0%	72.7%	70.3%	90.9%
	% of Total	4.7%	28.2%	30.6%	11.8%

The "within race" percentages in Table 22 did not give clear picture about the preference of non-white and white high school students' ratings of the survey item. The means of the non-white and white high school students' responses were very close, 2.66 and 2.62 respectively. With such a small difference, it was not surprising that the Mann-Whitney U test for comparing the non-white and white students did not indicate a significant result.

Intention to Work as STEM Professionals After College

The mean to the survey item "As a result of participating in the Forensic Challenge, I am inclined to continue to pursue a STEM professional career after I complete college" was 2.61 not indicating a very positive rating of the effects of the STEM Forensic Challenge on enhancing the high students' intention to work as STEM professionals after college. The frequency distribution of the non-white and white students' responses is presented in Table 23.

The "within race" percentages in Table 23 did not provide a clear picture about the preference of non-white and white high school students' ratings of the survey item: while the percentage of white high school students in the "disagree" category was much higher, there were no white high school students in the "strongly disagree" category but 9.4% of the non-white students in this category. The means of the non-white and white high school students' responses were very close, 2.63 and 2.57 respectively. With such a small difference, it was not surprising that the Mann-Whitney U test for comparing the non-white and white students' responses did not indicate a significant result.

Table 23. STEM Challenge: Intention to work as a STEM professional

Race	Reported Statistics	Strongly Disagree	Disagree	Agree	Strongly Agree
White	Count (n)	0	13	4	4
	% Within Race	0.0%	61.9%	19.0%	19.0%
	% Within Responses	0.0%	35.1%	15.4%	25.0%
	% of Total	0.0%	15.3%	4.7%	4.7%
Non-White	Count (n)	6	24	22	12
	% Within Race	9.4%	37.5%	34.4%	18.8%
	% Within Responses	100.0%	64.9%	84.6%	75.0%
	% of Total	7.1%	28.2%	25.9%	14.1%

STEM Challenge Survey: 2015

For the 2015 STEM Challenge, the high school students did a heart dissection for S (science), calculator powered robots for T (technology), a domino design for E (Engineering), and a math problem for M (mathematics) (see Table 2).

Interest in STEM Fields

The mean to the survey item "The STEM Challenge increased my interest in STEM fields" was 3.61 indicating a very positive rating of the effects of the STEM Challenge on increasing the high school students' interests in STEM fields. All survey participants were African American high school students. The frequency distribution of their responses is presented in Table 25.

Knowledge in STEM Fields

The mean to the survey item "The STEM Challenge increased my knowledge in STEM fields" was 3.70 indicating a very positive rating of the effects of the STEM Challenge on increasing the high school students' knowledge in STEM fields. All survey participants were African American high school students. The frequency distribution of their responses is presented in Table 26.

Table 24. Demographics of 2015 high school STEM Challenge survey participants

	2015 (n = 23)
Gender	Female: 11
	Male: 12
Ethnicity	Hispanic/Latino: 0
	Not Hispanic/Latino: 23
Race	African American: 23
	White: 0

Table 25. 2015 high school STEM Challenge: Interest in STEM fields

	Frequency	Relative Frequency
Agree	9	39%
Strongly Agree	14	61%

Table 26. 2015 high school STEM Challenge: Knowledge in STEM fields

	Frequency	Relative Frequency
Agree	7	30%
Strongly Agree	16	70%

Intention to Enroll as STEM Majors

The mean to the survey item "As a result of participating in the STEM Challenge, I am inclined to pursue a STEM discipline as a major in college" was 2.96 not indicating a very positive rating of the effects of the STEM Challenge on enhancing the high students' intention to enroll as STEM majors. All survey participants were African American high school students. The frequency distribution of their responses is presented in Table 27.

Intention to Work as STEM Professionals After College

The mean to the survey item "As a result of participating in the STEM Challenge, I am inclined to continue to pursue a STEM professional career after I complete college" was 2.87 not indicating a very positive rating of the effects of the STEM Forensic Challenge on enhancing the high students' intentions to work as STEM professionals after college. All survey participants were African American high school students. The frequency distribution of the responses is presented in Table 28.

Table 27. 2015 high school STEM Challenge: Intention to enroll as a STEM major

	Frequency	Relative Frequency
Strongly Disagree	1	4%
Disagree	3	13%
Agree	15	65%
Strongly Agree	4	17%

Table 28. 2015 high school STEM Challenge: Intention to work as STEM professional

	Frequency	Relative Frequency
Disagree	5	22%
Agree	16	69%
Strongly Agree	2	9%

SUMMARY AND FUTURE RESEARCH DIRECTIONS

The Career Panel Discussion survey collected data from the 2013 cohort of Project Engage undergraduate STEM students to assess the Career Panel Discussion activity's effects on increasing interest in STEM fields, knowledge in STEM fields, persistence as STEM majors, and intention to work as STEM professionals after graduation. The means of the responses to the four survey items were 3.33, 3.33, 3.44, and 3.44 indicating positive ratings of the effects of the Career Panel Discussion by the students. The sample size of the survey respondents was small and numbers of the non-white and white students were quite disproportionate. Future research is needed to evaluate the effects of similar activities on a larger number of undergraduate STEM students with more proportionate numbers of white and non-white students.

The analyses of the data collected from 2013, 2014, 2016 high school Forensic Challenge and the 2015 high school STEM Challenge focused on assessing the effects of the field trips on improving interest in STEM fields, knowledge of STEM fields, intention to enrolled as STEM majors, and intention to work as STEM professionals after college. The means for the high school students' responses to these four survey items are summarized in Table 29.

Table 29. Summary of high school STEM Challenge survey response means

	Means			
	Interest in STEM fields	Knowledge of STEM fields	Intention to enroll as STEM majors	Intention to work as STEM professionals after college
2013, 2014, 2016 Forensic Challenge	2.89	2.92	2.65	2.61
2015 STEM Challenge	3.61	3.70	2.96	2.87

As indicated by the means in the above table, the 2015 STEM Challenge received much higher ratings than 2013, 2014, and 2016 Forensic Challenge in terms of increasing high school students' interest in STEM fields and knowledge of STEM fields, but such a difference was not found in "intention to enroll as STEM majors" and "intention to work as STEM professionals after college." While the 2015 high school STEM Challenge included four different activities: heart dissection for S (science), calculator powered robots for T (technology), domino design for E (Engineering), and math problem for M (mathematics), the 2013, 2014, and 2016 Forensic Challenge only included the forensic crime scene investigation activity for all participating high school students. Does this explain why the 2015 STEM Challenge was received by the high school students as more effective in increasing interest and knowledge in STEM fields? If yes, why? These questions will serve as directions for future research efforts.

The analysis of the 2013, 2014, and 2016 STEM Forensic Challenge data identified significantly higher ratings of the Forensic Challenge in regard to the effect on increasing the knowledge of STEM fields by non-white high school students. The significant achievement gaps among white and non-white or minority student populations are well documented in the literature (Ingels, Pratt, Rogers, Siegel, & Stutts, 2004; National Research Council, 2011; Riegle-Crumb & Grodsky, 2010). Are such achievement gaps related to the above mentioned difference between white and non-white students? Future research is needed to help shed light on this question and to help make decisions about designing and developing effective STEM activities or projects to improve rural non-white or minority students' knowledge of STEM fields.

CONCLUSION

Career exploration and career self-efficacy dictate student career interests (Turner & Lapan, 2002); therefore, it is necessary to introduce students to STEM careers early through outreach programs and prepare them for the STEM workforce during their college years. A tripartite approach to career exploration develops career decidedness, positive STEM career thoughts, and ultimately vocational maturity. Project Engage accomplishes the three tiers by first, introducing freshmen to STEM careers with a course focused solely on STEM careers, assisting students with matching their interests and personalities with the most appropriate STEM field (career decidedness). Secondly, the career exploration course supports STEM students growing vocational maturity as students are given practice in resume writing, and interviewing, as well as opportunities for researching STEM careers. Lastly, STEM guest speakers, career panelists, networking, and research opportunities foster the development of STEM self-efficacy and dispel negative STEM career thoughts. Collectively, these activities constitute a tripartite plan for guiding students toward STEM careers.

However, the growth of career thoughts should not begin during the freshman year of college. Adolescents begin to lose interest in STEM between 5th and 8th grade (Vedder-Weiss & Fortus, 2011; 2012); therefore, there is a need to capture STEM interest in the early grades and sustain STEM interest and STEM career interest throughout the secondary level. Project Engage conducts outreach to rural schools offering STEM competitive events (STEM Challenge and Invention Convention) which serve to foster interest in STEM careers. Survey results indicate the events created positive STEM thoughts among high school students.

Project Engage seeks to increase the STEM workforce numbers not only by recruiting and retaining rural students in STEM majors, but also by preparing them to make informed decisions regarding STEM career choices. By implementing a tripartite approach to career exploration, Project Engage leads students to improved decision-making, positive career thoughts. and increased awareness of STEM career fields.

ACKNOWLEDGMENT

This research was supported by the U. S. Department of Education Minority Science and Engineering Program [grant number P120A120099] and the U. S. Department of Education Minority Science and Engineering Program [grant number P120A160122].

REFERENCES

Arnett, J. J. (2007). *International encyclopedia of adolescence*. New York: Routledge.

Ates, O., & Eryilmaz, A. (2011). Effectiveness of hands-on and minds-on activities on students' achievement and attitudes towards physics. *Asia-Pacific Forum On Science Learning and Teaching*, *12*(1), 1–22.

Bigler, A. M., & Hanegan, N. L. (2011). Student content knowledge increases after participation in a hands-on biotechnology intervention. *Journal of Science Education and Technology*, *20*(3), 246–257. doi:10.100710956-010-9250-7

Blankenburg, J. S., Höffler, T. N., & Parchmann, I. (2016). Fostering today what is needed tomorrow: Investigating students' interest in science. *Science Education*, *100*(2), 364–391. doi:10.1002ce.21204

Caglak, S. (2017). Does hands-on science practices make an ımpact on achievement in science? A meta-analysis. *Journal of Education in Science Environmental Health*, *3*(1), 69–87. PMID:28637469

Carpi, A., Ronan, D. M., Falconer, H. M., & Lents, N. H. (2017). Cultivating minority scientists: Undergraduate research increases self-efficacy and career ambitions for underrepresented students in STEM. *Journal of Research in Science Teaching*, *54*(2), 169–194. doi:10.1002/tea.21341

Chason, A. K., Bullock-Yowell, E., Sampson, J. P. Jr, Lenz, J. G., & Reardon, R. C. (2013). Relationships among career thoughts, career interests, and career decision state. *The Canadian Journal of Career Development*, *12*(1), 41–47.

Dillman, D. A. (2006). Why choice of survey mode makes a difference. *Public Health Reports (1974-)*, *121*(1), 11-13.

Gnilka, P. B., & Novakovic, A. (2017). Gender differences in STEM students' perfectionism, career search self-efficacy, and perception of career barriers. *Journal of Counseling and Development*, *95*(1), 56–66. doi:10.1002/jcad.12117

Holstermann, N., Grube, D., & Bögeholz, S. (2009). Hands-on activities and their influence on students' interest. *Research in Science Education*, *40*(5), 743–757. doi:10.100711165-009-9142-0

Ingels, S. J., Pratt, D. J., Rogers, J. E., Siegel, P. H., & Stutts, E. S. (2004). *Education longitudinal study of 2002: Base-year data file user's manual (NCES 2004-405)*. Washington, DC: National Center for Education Statistics, Institute of Education Sciences, U.S. Department of Education.

Lounsbury, J. W., Saudargas, R. A., & Gibson, L. W. (2004). An investigation of personality traits in relation to intention to withdraw from college. *Journal of College Student Development*, *45*, 517–534.

Marle, P. D., Decker, L., Taylor, V., Fitzpatrick, K., Khaliqi, D., Owens, J. E., & Henry, R. M. (2014). CSI-Chocolate science investigation and the case of the recipe rip-off: Using an extended problem-based scenario to enhance high school students' science engagement. *Journal of Chemical Education, 91*(3), 345–350. doi:10.1021/ed3001123

McDonald, C. V. (2016). STEM Education: A review of the contribution of the disciplines of science, technology, engineering and mathematics. *Science Education International, 27*(4), 530–569.

National Research Council. (2011). *Successful K-12 STEM education: Identifying effective approaches in science, Technology, Engineering and Mathematics*. Washington, DC: The National Academies of Press. Retrieved from http://nap.edu/13158

Netemeyer, R. G., Bearden, W. O., & Sharma, S. (2003). *Scaling procedures: Issues and applications*. Thousand Oaks, CA: Sage Publications. doi:10.4135/9781412985772

Paivandy, S., Bullock, E. E., Reardon, R. C., & Kelly, F. D. (2008). The effects of decision-making style and cognitive thought patterns on negative career thoughts. *Journal of Career Assessment, 16*(4), 474–488. doi:10.1177/1069072708318904

Peterson, G. W., Sampson, J., & Reardon, R. (1991). *Career development and services: A cognitive approach*. Pacific Grove, CA: Brooks/Cole.

Prescod, D. (2014). *The influence of a career planning STEM explorations course on vocational maturity, career decidedness, and career thoughts for undergraduate students* (Doctoral dissertation). Retrieved from http://stars.library.ucf.edu/cgi/viewcontent.cgi?article=4041&context=etd

Riegle-Crumb, C., & Grodsky, E. (2010). Racial ethnic differences at the intersection of math course-taking and achievement. *Sociology of Education, 83*(3), 248–270. doi:10.1177/0038040710375689

Sampson, J. P. Jr, Reardon, R. C., Peterson, G. W., & Lenz, J. G. (2004). *Career counseling and services: A cognitive approach*. Pacific Grove, CA: Wadsworth-Brooks/Cole.

Savickas, M. L., Briddick, W. C., & Watkins, C. E. Jr. (2002). The relation of career maturity to personality type and social adjustment. *Journal of Career Assessment, 10*(1), 24–49. doi:10.1177/106907270201000l002

Schneider, B., Broda, M., Judy, J., & Burkander, K. (2013). Pathways to college and STEM careers: Enhancing the high school experience. *New Directions for Youth Development, 2013*(4), 9–29. doi:10.1002/yd.20076 PMID:24474256

Siew, N. M., Goh, H., & Sulaiman, F. (2016). Integrating STEM in an engineering design process: The learning experience of rural secondary school students in an outreach challenge. *Journal of Baltic Science Education, 15*(4), 477–493.

Sisters of the Academy. (2015). *Priming the pipeline*. Retrieved from http://www.sistersoftheacademy.org/page/PrimingthePipeline

Smith, R. M. (2011). *Personality traits and career decidedness: An empirical study of university students* (Doctoral dissertation). Retrieved from: http://trace.tennessee.edu/cgi/viewcontent.cgi?article=2074&context=utk_graddiss

Turner, S., & Lapan, R. T. (2002). Career self-efficacy and perceptions of parent support in adolescent career development. *The Career Development Quarterly, 51*(1), 44–55. doi:10.1002/j.2161-0045.2002.tb00591.x

Vedder-Weiss, D., & Fortus, D. (2011). Adolescents' declining motivation to learn science: Inevitable or not? *Journal of Research in Science Teaching, 48*(2), 199–216. doi:10.1002/tea.20398

Vedder-Weiss, D., & Fortus, D. (2012). Adolescents' declining motivation to learn science: A follow-up study. *Journal of Research in Science Teaching, 49*(9), 1057–1095. doi:10.1002/tea.21049

Wang, M.-T., & Degol, J. (2013). Motivational pathways to STEM career choices: Using expectancy-value perspective to understand individual and gender differences in STEM fields. *Developmental Review, 33*(4), 304–340. doi:10.1016/j.dr.2013.08.001 PMID:24298199

Yadav, A., Stephenson, C., & Hai, H. (2017). Computational thinking for teacher education. *Communications of the ACM, 60*(4), 55–62. doi:10.1145/2994591

KEY TERMS AND DEFINITIONS

Career Decidedness: The degree to which individuals feel certain about their career choice.

Career Self-Efficacy: Beliefs regarding their abilities to perform career behaviors in relation to career choice.

Career Thoughts: Feelings, assumptions, attitudes, beliefs, behaviors, expectations, and plans or strategies related to career decision making.

Guided Career Exploration: Activities which inform undergraduate STEM majors of their career choices in STEM.

High School Outreach: Activities provided by Project Engage for high school students and teachers.

Invention Convention: Supports student creativity and problem-solving. Students design, build and test inventions providing a real-world STEM experience.

Sisters of the Academy: An educational network of Black women in higher education.

STEM Challenge: A competitive event which involves students using skills in science, technology, engineering, and mathematics.

Vocational Maturity: The ability level of an individual to make informed career decisions.

Chapter 6
Improving Undergraduate STEM Education:
A Four-Dimensional Framework

ABSTRACT

Improving STEM education in the U.S. stands to address one of the nation's priorities, increasing the STEM workforce. Rural students are a previously untapped source of potential; yet, these students lack STEM experiences, engagement, and scaffolding to stay motivated to complete an undergraduate program in STEM. Motivation is a key element for rural student interest and persistence in STEM; however, these students also must develop certain knowledge and skills to succeed as STEM professionals. The authors propose a model, Improving Undergraduate STEM Education Framework, as a means to maintain motivation and prepare rural students for the rigors of a STEM major/career. In this chapter, the four-dimensional framework is unveiled along with examples of Project Engage's efforts to address each dimension.

INTRODUCTION

Through the support of a U. S. Department of Education Minority Science and Engineering Program grant, Project Engage programs and activities were designed and applied to achieve the ultimate goal of improving rural undergraduate STEM programs' capacity of retaining undergraduate students in STEM majors, cultivating STEM talents, and producing motivated and

DOI: 10.4018/978-1-5225-6341-9.ch006

dedicated STEM graduates. These programs and activities introduced in previous chapters include a Strategic Mentoring Program, Professional Exploration Trips, Competitive STEM Events, Invited STEM Speakers, Team Building Training, Career Exploration Course, Career Panel Discussion, Undergraduate Research, Sisters of the Academy "Priming the Pipeline," High School Student Outreach, Invention Convention and STEM Challenge. Based on its experience in these programs and activities, Project Engage developed a four-dimensional practice-and research-based framework that can be used to guide future efforts of improving undergraduate STEM education in rural areas. The four dimensions of this Improving Undergraduate STEM Education Framework are: motivating, experiencing, engaging, and scaffolding (See Figure 1).

THE "LEAKY" PIPELINE

Using the analogy of a "STEM pipeline" representing the pathway from K-12 education to college STEM majors to future STEM careers, recent efforts have focused on increasing the flow of students through the pipeline.

Figure 1. The Improving Undergraduate STEM Education Framework

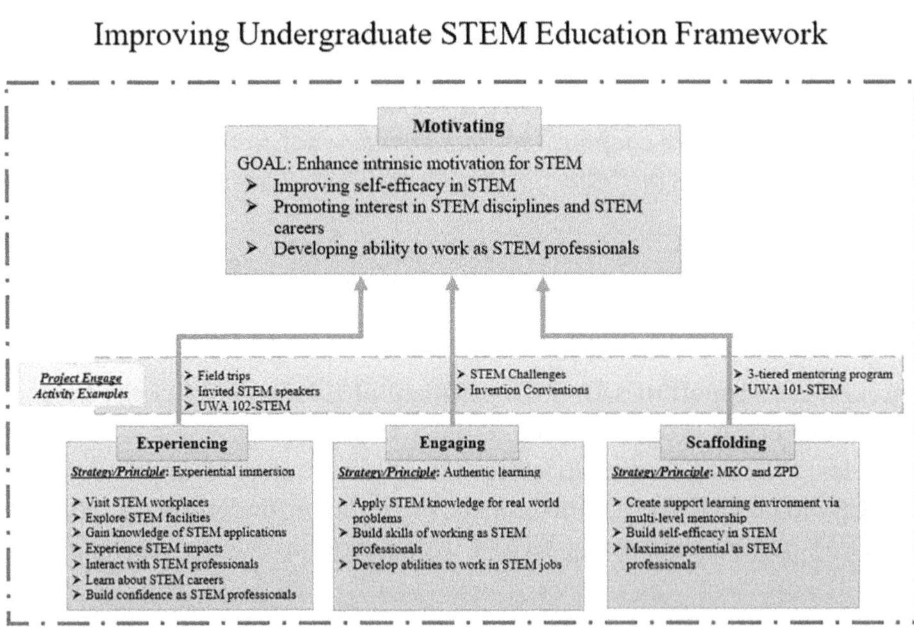

Compounding the problem of increasing the numbers of students going *into* the STEM pipeline are "leaks" *out of* the pipeline. Current research has identified such "leaks" including: 1) students increasingly losing interest in STEM as they matriculate from elementary to middle to high school; 2) students who graduate from college with a STEM degree, but find employment in another field; 3) unfavorable working environments in STEM for groups such as minorities and females, and; 4) lack of motivation to preserve in STEM. Leaks in the STEM pipeline result in a lowered flow of students out of college STEM fields and into the STEM workforce (Versypt & Versypt, 2013). There is a need to increase the flow into the STEM pipeline, especially for diverse and underrepresented groups. Rural school districts represent more than half of all operating regular school districts (National Center for Education Statistics, 2013). Due to the large number of rural schools, focus should be turned towards rural students, an underrepresented group, as an untapped source of students to increase the supply of students into the STEM pipeline (Versypt & Versypt, 2013).

Attrition rates for undergraduate students in STEM areas remain high (Tinto, 1993; Wilson et al, 2012) and less than 50% of the undergraduate students entering as STEM majors are actually graduating with a STEM degree (Hayes, Whalen & Cannon, 2009; Wilson et al., 2012). Black STEM students are even less likely to persist in STEM majors than white students (Flynn, 2016). The situation in rural areas is even worse. Educational environments in rural areas are characterized by: high levels of poverty; less available cultural capital and consequent lack of enrichment activities outside of the regular school curriculum; lack of qualified STEM teachers; lack of role models of successful STEM professionals; and inadequate awareness of STEM career opportunities. Growing up in such an educational environment, rural STEM majors are not well prepared for academic success in the learning of their STEM disciplines and lack motivation to complete their STEM degree and pursue a STEM career after graduation. Increasing the retention of rural students in STEM majors and preparing them for STEM careers has a direct and positive impact by increasing the potential for future financial earnings, but also directly benefits the nation two-fold: by supplying the STEM pipeline with a new source of "flow" and increasing the diversity of the STEM workforce (Flynn, 2016). Project Engage's four dimensional Improving Undergraduate STEM Education Framework, adopting a comprehensive approach to STEM education, was developed to address the needs of rural students, particularly rural minority students in support of STEM degrees and ultimately preparation for STEM careers.

MOTIVATION: THE KEY TO IMPROVED STEM EDUCATION

Motivation to pursue STEM degrees and careers is the key to improving undergraduate STEM education in rural areas. Many times, students choose to pursue STEM majors in college initially, but change their majors before graduation. This can be attributed in part to a lack of motivation due to perceptions of difficulty and/or uninteresting STEM courses (Rosenzweig & Wigfield, 2016). Intrinsic motivation is the self-desire to seek out new things and new challenges, to analyze one's capacity, to observe and to gain knowledge (Ryan & Deci, 2000). Intrinsic motivation is driven by one's interest or enjoyment in a particular task, and it stimulates and directs goal oriented behaviors. Intrinsically motivated students are more likely to engage willingly in a task and work to improve their skills. Rural students have historically had slightly lower motivation to pursue professional STEM careers than their urban peers. The lower aspirations have been attributed to both socioeconomic differences as well as disparate occupational desires between rural and their nonrural counterparts (Haller & Virkler, 1993). Interventions designed to increase motivation are needed to improve educational and career outcomes for rural students. The *motivating* dimension of the Improving Undergraduate STEM Education Framework centers on improving undergraduate STEM students' intrinsic motivation through the following aspects: improving undergraduate students' self-efficacy in STEM learning; increasing their interest in STEM disciplines and STEM careers; and enhancing their ability to work as STEM professionals.

STEM Self-Efficacy

A student's belief in his/her knowledge and skills related to success in science, technology, engineering and/or mathematics is student STEM self-efficacy. A student's self-efficacy beliefs contributes to the individual's academic abilities (Bandura, 1993), goals an individual establishes, motivation to persist in obtaining these goals (Bandura 1997; Pajares, 2005; Bandura & Locke, 2003), and overall accomplishments (Bandura, 1993). According to Bandura and Locke, "…efficacy beliefs contribute significantly to the level of motivation and performance" (2003, p.87). Similarly, Tang, Pan, and Newmeyer (2008) concluded that there exists a strong correlation between career self-efficacy and career interest. Thus, STEM self-efficacy needs to start early through outreach to K-12 schools and continue throughout the college career. STEM self-efficacy can be fostered through authentic learning experiences involving critical thinking and problem solving skills (McDonald, 2016). Project Engage provided outreach to local schools by working with local rural STEM teachers and students in competitive STEM events. K-12 students built self-efficacy as they successfully competed as part of a team in authentic learning experiences. Competitive events, such as the Invention Convention and STEM Challenge, gave students opportunities to display their knowledge and skills in STEM. By building STEM self-efficacy in K-12 students, Project Engage was investing in recruitment of future STEM undergraduates. Moreover, a primary objective of Project Engage is to ensure student STEM self-efficacy continues to develop throughout the college career. The Strategic Mentoring Program supported STEM freshmen self-efficacy improvement through academic, social and general mentoring. Learning from older peer mentors, the career counselor, and STEM faculty helped students build confidence in their STEM majors. Competitive STEM events such as robotics, rover building, and boomerang design gave students the opportunity to have success in developing and displaying STEM knowledge and skills. Success in these areas motivated students to continue in STEM majors by supporting beliefs in their own abilities.

Interest in STEM Careers

Encouraging interest in STEM careers can be accomplished with a multi-faceted approach. A course specific to STEM career explorations, a career panel discussion board, guest speakers, and a career counselor were all features of Project Engage's efforts to motivate student interest in STEM careers. However, the crucial aspect introducing students to the endless possibilities of working in STEM was the professional exploration field trips.

Discovering the multiplicity of STEM careers supports the *Motivating* dimension of the Improving STEM Education Framework by opening students' eyes to the diversity of STEM careers. A course focused on STEM careers, assessment of an individual's proclivity for a career in STEM, as well as the development of skills needed for career entry was implemented during the second year of Project Engage to expand undergraduates' concept of STEM careers. STEM 102 Career Explorations provided an intensive examination of STEM careers. In addition, students were given personality tests, personal interest inventories, opportunities to attend career fairs, career panel discussions and more. The course served not only for career exploration activities, but also as a means to maintain the cohesiveness of the freshmen cohort.

Interaction with STEM professionals also allows students to learn about STEM careers and to build self-confidence of working as STEM professionals. Exposure to STEM role models has positive effects on STEM students' interest in STEM and academic self-efficacy (Shin, Levy, & London, 2016), yet students from rural areas are not often exposed to STEM role models. STEM Role models are hypothesized to increase female self-confidence (Betz & Hackett, 1981) and, thereby, positively impact female career aspirations. Project Engage afforded students opportunities to interact with and learn from STEM professionals through professional exploration field trips and invited STEM speakers.

Guest speakers from a variety of STEM professions and representative of the students' cultural and ethnic backgrounds provided insights into a multitude of STEM careers. Students were able to identify with the speakers who were carefully chosen so that Project Engage students were able to visualize their capacity for achieving success in STEM. Having STEM role models undoubtedly increases identity compatibility between the students and STEM (Shin et al., 2016). Participants in Project Engage were from the rural Black Belt region of Alabama, characterized by high poverty and low economic and educational performance. These students were in need of STEM role models who have also overcome such barriers to excel in STEM.

Professional exploration field trips, which were specifically designed with career exploration in mind, permitted students to learn about STEM careers (educational requirements, market demand, internship opportunities, etc.) directly from the STEM expert in the field. For example, Project Engage's trip to the Alabama Nature Center was designed so that students rotate through stations on the wildlife preserve where they interacted with a biologist, chemist or environmental scientist while he/she was working. Students witnessed the scientists performing their jobs all the while asking questions about the career. It is one thing for students to sit and listen to an instructor discuss various careers and requirements, but to hear first-hand from the STEM professional in the work setting is a powerful motivating force.

Developing the Ability to Work as STEM Professionals

STEM professionals must possess specific skills which can be classified into two categories: hard and soft; yet, "STEM skills are problematic to define as they do not exist in isolation; similar to innovation skills they are guiding, enabling or facilitating skills and borrow content from other skills groups"

(Siekmann & Korbel, 2016, p. 6). Hard skills such as critical thinking, problem-solving, argumentation, and mathematical skills are necessary across STEM disciplines.

Minnesota State Careerwise Education (2017) identified hard skills as listed here.

- *Analytical skills to research a topic, develop a project plan and timeline, and draw conclusions from research results.*
- *Science skills to break down a complex scientific system into smaller parts, recognize cause and effect relationships, and defend opinions using facts.*
- *Mathematic skills for calculations and measurements.*
- *Attention to detail to follow a standard blueprint, record data accurately, or write instructions.*
- *Technical skills to troubleshoot the source of a problem, repair a machine or debug an operating system, and computer capabilities to stay current on appropriate software and equipment.*

Through Project Engage, STEM freshmen were introduced to and developed STEM skills through STEM competitive events such as robotics. Robotics requires the use of design skills as well as coding ability. For some Project Engage students, especially those from more deprived backgrounds, working with robotics was completely foreign. Through intensive training, students developed their abilities: problem-solving/analytical, how to make use of the materials provided; science/mathematics, researching and constructing robots with intricate parts; and technical/logical reasoning skills, such as programming robots that would perform specified tasks.

Soft skills such as creativity, communication, leadership, also play into the ability to work as a STEM professional. These skills facilitate teamwork that has become an important concept in today's STEM workforce with STEM projects no longer individual, but requiring group efforts from individuals with varying skill sets. Soft skills as recognized by Minnesota State Careerwise Education (2017) are included in the following list.

- *Communication and cooperation skills to listen to customer needs or interact with project partners.*
- *Creative abilities to solve problems and develop new ideas.*
- *Leadership skills to lead projects or help customers.*
- *Organization skills to keep track of lots of different information.*

Project Engage students were encouraged to develop soft skills through a variety of activities such as the design and building challenges (robotics, hover craft, boomerangs) or team building activities for developing communication, collaboration and leadership skills (e.g., 4 H Center activities, in class teamwork instruction, leadership in the high school outreach activities).

The ability to work as a STEM professional is complex and requires knowledge and skills that must be nurtured as a student progresses through college. Rural students often have the innate soft skills needed for success in STEM, but need assistance developing those hard, technical skills to which they may not have been exposed during their secondary education. Project Engage activities and services supported the development of such skills and in doing so strengthened the first dimension of the Improving STEM Education Framework, *motivating* students to persevere in STEM majors and ultimately reinforcing the STEM pipeline.

EXPERIENCING STEM

The other three dimensions (i.e., *Experiencing*, *Engaging*, and *Scaffolding*) in the Improving Undergraduate STEM Education Framework serve as means to achieve the ends of the *Motivating* dimension. The *Experiencing* dimension focuses on learning through experience. This learning through experience can be achieved through an experiential immersion strategy of providing undergraduate students opportunities to: 1) visit STEM workplaces to explore STEM facilities and gain personal understanding of real world applications, 2) interact with STEM professionals to learn about STEM careers and to build self-confidence of working as STEM professionals, and 3) experience STEM impacts to promote interest in STEM. Students from rural locales typically are not exposed to rich STEM experiences or STEM professionals. Key to this experiential immersion is giving students substantial exposure to and personal experience with real world STEM.

The *Experiencing* dimension of the Improving Undergraduate STEM Education Framework can be accomplished through visits to STEM workplaces. Professional exploration field trips furnish students exposure to real world applications of STEM and STEM professionals simultaneously. Rural students may only have a textbook perspective of what STEM is; therefore, exposing them to STEM industries can expand their vision of STEM and STEM careers. According to Strimel (2014), students are not often exposed to authentic learning experiences, but are limited to content set forth by the school/teacher or dictated by assessment purposes. During onsite visits to STEM facilities, students observe real world applications of the math and science they learn in the classroom resulting in an increased understanding of the impact of STEM. Project Engage provided field trips to STEM industries such as the Marshall Space Flight Center, Alabama Nature Center, and HudsonAlpha Institute for Biotechnology for STEM undergraduates. By providing an assortment of STEM facilities for the field trips (aeronautical engineering, environmental science, genomics, etc.), students were able to experience a STEM field of interest or perhaps discover a new area of interest. Activities conducted on the STEM industry sites, such as robotics, rover building, water testing, seining for invertebrates, helped the students build the confidence that they too can work as a STEM professional. Witnessing STEM in action and "doing" STEM are exciting experiences that can increase a student's motivation to pursue or persevere in a STEM major.

Kassaee and Rowell (2016) found that guest speakers can speak about career opportunities in STEM for research and development projects to motivate students to persist in their STEM coursework. Project Engage sought out STEM professionals who had similar backgrounds as the student population and high levels of success in their fields to bring to campus as invited speakers. Being a highly successful STEM professional with a similar background or racial identity was important for establishing the speaker as a role model for Project Engage students. Relating to the personal stories and struggles of the speakers, seeing the success of the speakers as something that they too can achieve, and experiencing the anecdotal accounts directly from the speakers inspired Project Engage students to see their own potential in STEM.

ENGAGING IN STEM

Engagement theory states, "…it is the individual's integration into the academic and social systems of the college that most directly related to his continuance in that college" (Tinto, 1993). The *Engaging* dimension centers on the authentic learning strategy of engaging students in hands-on activities of solving real world STEM problems. Such hands-on activities entail the application of STEM knowledge as well as skills and abilities required of STEM professionals. Therefore, these hands-on activities are intended to help undergraduate students increase their knowledge and abilities of STEM applications, and build their skills and abilities of working as STEM professionals. To achieve these intended outcomes, these hands-on activities can be designed and developed using the four types of authenticity (i.e., context authenticity, task authenticity, impact authenticity, and personal/value authenticity) (Strobel, Wang, Weber, & Dyehouse, 2013) as guidance. The four types of authenticity and their definitions are listed in Table 1.

Contextual Authenticity

Project Engage employed robotics, rover building, and forensics events to simulate real world situations. Contextual authenticity or activities steeped in real-world situations motivate student learning and promote retention in the STEM major. "This type of authenticity should take place in authentic contexts and resemble daily life experiences" (Strobel et al., 2013, p. 146). Robotics began as an activity accomplished through the STEM-101 freshmen seminar course but due to the time constraints imposed by completing the robotics program with an existing course and the lack of an authentic work environment, robotics was moved offsite to the Marshall Space Flight Center. At the Center, students received first-hand instruction from robotics experts and worked in an environment simulating an authentic research lab. Likewise, the rover building activity was also held on site at the Marshall Space Flight Center. The scientists at the Center were able to provide an environment that simulated the working environment of an aeronautical engineer. Project Engage also used forensic activities that were designed to simulate actual

Table 1. Four types of authenticity

Categories	Description/Key Definition
Contextual Authenticity	1. Contains a real-world professional environment 2. Provides everyday experience 3. Resemble social challenges or social interactions 4. Align school and professional outcomes
Task Authenticity	1. Students are engaged in tasks similar to professionals 2. Reflect or develop professional skills for students 3. Challenge students in decision-making in practical contexts
Impact Authenticity	1. Actions in cultural practices 2. Social events, issues or impact 3. Students' role as effective citizens 4. Students as collaborators of industry
Personal/Value Authenticity	1. Produce knowledge with values in students' lives and studies beyond simply proving their competence 2. Integrate everyday life experience, personal interests, professional target and cultural values 3. Develop a sense of identity and sense of confidence

(Strobel et al., 2013)

crime scene investigations. Students were given the backstory of a fictitious crime and were then required to go to various crime scenes and collect evidence. Evidence had to be transported to the "crime lab," an actual lab on campus, for analysis. The use of simulated crime scenes and actual labs created an authentic context.

Task Authenticity

According to Strobel et al. (2013), task authenticity "...focuses on constructivist type learning environments in which students may be challenged to make decisions in practical contexts" (p. 146-147). Task authenticity engages students in tasks similar to those of a STEM professional. Robotics activities require students to engage in design, building, coding and then redesign in order to achieve a finished product capable of completing a pre-determined task. Project Engage participants were challenged with learning the knowledge and skills required for creating robots. These tasks are realistic, simulating research and development of robots for the workforce. Likewise, building mock space rovers simulates a task of a research scientists/engineer. Students were required to engage in creating a rover design, building the model, testing, and making needed modifications. The process simulated the authentic task of a research scientist or engineer working on a spaceflight craft. The forensics STEM Challenge required students to collect, analyze and report the findings on a crime scene investigation. This authentic task incorporated all areas of STEM, engaging students in a task that has been popularized on television. As a result, students built their knowledge and abilities of STEM applications, and improved their skills and abilities of working as STEM professionals.

Impact Authenticity

The concept of impact authenticity places students into social issues where they play a role as an effective STEM professional. According to Barab, Squire, and Dueber (2000), impact authenticity goes even further to include products of students being utilized outside of school. Project Engage's outreach to high school students provided opportunities to explore societal issues. For example, students from local rural schools investigated recycling when they participated in a "Green Convention." As part of this activity students had to design a useful product from discarded trash. This activity reinforced the benefits of recycling for society. Another example of impact authenticity required students to research population growth data in our local area, determine if there was a potential problem, and design a plan to address water usage issues. Through this authentic task, students were able to make a real connection with a local social issue.

Personal/Value Authenticity

Tasks which are perceived to have personal authenticity produce knowledge beyond competence, knowledge that has personal meaning or value to the student (Strobel et al., 2013). For example, student interactions with STEM professionals during professional field trips to gain knowledge regarding career interests have personal authenticity. The information gained does increase academic knowledge, but it can be invaluable to a student's motivation to persist on a STEM pathway. Additionally, developing professional e-portfolios is an authentic task that can have a great impact on student personal success in pursuing a career. Student activities and achievements such as career inventories, personality tests, service learning experiences, internships, work experience and more constitute actions, when combined into an e-portfolio, paint a complete picture of a student's academic development which can be shared with future employers. Project Engage promoted the development of such portfolios through the aid of the career counselor. The career counselor began working with students to develop a professional portfolio in their sophomore year and continued to completion. The development of the portfolio is an important personal authentic task.

Within the Improving STEM Education Framework, the *Engaging* dimension demands more authentic, hands-on, discovery based activities to engage students in STEM learning. Meaningful tasks motivate students to develop critical-thinking and problem-solving skills (Strimel, 2014). Engaging students in more active STEM learning motivates students to persist through college and into the STEM workforce pipeline.

SCAFFOLDING

The *Scaffolding* dimension applies the MKO (More Knowledgeable Other) and the ZPD (Zone of Proximal Development) principles to create a supportive learning environment for undergraduate STEM students. In this supportive learning environment, scaffolds can come in the form of mentorship between undergraduate STEM students and mentors of three different levels: upperclassmen mentors, graduate level career counselors, and STEM faculty mentors. Interactions with peer mentors, a career counselor and faculty mentors serve to enhance student self-efficacy beliefs. Project Engage placed college STEM students in supportive, nurturing learning environments by submersing them in a tripartite scaffolded mentoring system involving peer mentors, a career counselor, and faculty STEM mentors. This supportive learning environment gave rural students, specifically those first-generation students, an opportunity to develop STEM self-efficacy in a non-threatening manner. Peer mentors provided insight into academic course offerings, college social life, tutoring, and more. A career counselor provided a more advanced role model who focused on disseminating career information. The third level of mentoring was delivered by STEM faculty through interactions within the STEM 101 freshmen seminar class as well as outside the class through professional social interaction such as STEM clubs and laboratory research. The mentors of the three levels served as more knowledgeable others for undergraduate STEM students to allow them to build self-efficacy in STEM, reach a higher level of development (Zone of Proximal Development), and maximize their potential as STEM professionals.

SUMMARY AND FUTURE RESEARCH DIRECTIONS

The *Experiencing*, *Engaging*, and *Scaffolding* dimensions in the Improving Undergraduate STEM Education Framework emerged from the programs and activities designed for and implemented among Project Engage students. Survey were conducted among the students to evaluate the impact of Project Engage program and activity. The surveys were program- or activity-specific. Due to the differences in the characteristics and purposes of the programs and activities, each survey focused on different aspects in evaluating the impact. The following is a summary of these aspects.

1. Overall program/activity effect
2. Competency to be successful in STEM
3. Knowledge of STEM fields
4. Persistence as STEM majors
5. Intention to work as STEM professionals after graduation
6. Interest in STEM fields
7. Team cooperation

8. Personal confidence
9. Team effectiveness strategies

The survey data analysis results reported in previous chapters not only indicate overall positive effects of Project Engage programs and activities but also positive impact in terms of the aspects listed above. Comparing the aspects with the goals in the *Motivating* dimension in Figure 1 will lead to the following associations, which indicate that Project Engage programs and activities aligned well with the goal and sub-goals presented in the *Motivating* dimension.

Goal: Enhance Intrinsic Motivation for STEM

- Improving self-efficacy in STEM
  - Competency to be successful in STEM
  - Knowledge of STEM fields
- Promoting interest in STEM disciplines and STEM careers
  - Persistence as STEM majors
  - Intention to work as STEM professionals after graduation
  - Interest in STEM fields
- Developing ability work as STEM professionals
  - Team cooperation
  - Personal confidence
  - Team effectiveness strategies

The revisit of the survey data analysis results as discussed above, together with the detailed descriptions of Project Engage programs and activities provided in chapters III-V indicate that the Improving Undergraduate STEM Education Framework emerged from and is solidly grounded in the practices and research of Project Engage. This means that this practice- and research-based framework can serve as a valid and useful tool guiding STEM educators and STEM educational researchers in their design and development of future interventions or initiatives aimed to improve undergraduate STEM education in rural areas. However, STEM educators and STEM educational researchers need to be aware of the fact that the Improving Undergraduate STEM Education Framework was developed based on STEM intervention practices and research applied to Project Engage undergraduate STEM students in a specific area of the U. S. known as the Black Belt region. These students and the Black Belt region itself have unique characteristics that may not be found in other rural areas. Therefore, modifications and improvements are unavoidable when the Improving Undergraduate STEM Education Framework is used on undergraduate STEM students in other rural areas. It is envisioned that future efforts and research intended to promote STEM student success in other rural areas will provide practical experience and empirical evidence to allow the framework to be further developed into a tool for improving STEM education across rural areas in the nation.

CONCLUSION

Rural students are a potential source for the STEM pipeline, but efforts have to be made to support their specific needs as they 1) prepare for college during the K-12 years, 2) transition from high school to a college as a STEM major, and 3) progress through college so that they persist as a STEM major. The proposed Improving Undergraduate STEM Education Framework holds promise for addressing the needs of rural students in order for them to enter the STEM pipeline and not "leak" out before entering the STEM workforce. Rural students first have to be introduced to STEM careers as they often have limited views of potential careers. Once introduced to the many possible STEM careers, these students have to be motivated to pursue and persist in a STEM major. Due to limited experiences with STEM careers and STEM professionals, it is vital to provide the experiences needed to develop knowledge and skills needed for working as a STEM professional. Furthermore, rural students require authentic STEM learning experiences to stay engaged. Without these authentic experiences, students may lose focus and choose a less rigorous major. Finally, by providing scaffolding where the rural students receive support and learn from mentors (More Knowledgeable Others), these students can be led to a higher level of understanding of STEM (Zone of Proximal Development). The time has come for science, technology, engineering, and mathematics learning to be experiential and authentic, no longer restricted to the four walls of a classroom.

The comprehensive approach as illustrated by the four dimensional framework (Figure 1) successfully addressed the needs of rural students and improved their knowledge and competence in STEM, interest in STEM disciplines and careers, and abilities to work as STEM professionals. The positive effects, collectively, rendered a positive impact on Project Engage STEM students' retention rate. Project Engage efforts were concentrated on the first year of college for rural STEM students. The retention data of

Project Engage students, freshmen to sophomore year, indicated that Project Engage was positively impacting retention of first year STEM students (45%, 54.10%, 80.50%, 59.50% for cohorts 2012-2015). Likewise, retention data over an extended time has improved over the baseline data established prior to the implementation of Project Engage. The first Project Engage cohort (2012) had a retention rate of 23% over four years compared to the 2010 cohort four-year retention rate of 12%. The Improving Undergraduate STEM Education Framework depicts a comprehensive support system that increases the flow of workers into the STEM pipeline and eliminates some of the so called "leaks."

ACKNOWLEDGMENT

This research was supported by the U. S. Department of Education Minority Science and Engineering Program [grant number P120A120099] and the U. S. Department of Education Minority Science and Engineering Program [grant number P120A160122].

REFERENCES

Bandura, A. (1993). Perceived Self-Efficacy in cognitive development and functioning. *Educational Psychologist*, *28*(2), 117–148. doi:10.120715326985ep2802_3

Bandura, A. (1997). Self-efficacy. *The Harvard Mental Health Letter*, *13*(9), 4.

Bandura, A., & Locke, E. A. (2003). Negative self-efficacy and goal effects revisited. *The Journal of Applied Psychology*, *88*(1), 87–99. doi:10.1037/0021-9010.88.1.87 PMID:12675397

(2000). Barab, Squire, & Dueber (2000). A coevolutionary model for supporting the emergence of authenticity. *Educational Technology Research and Development*, *48*(2), 37–62. doi:10.1007/BF02313400

Betz, N. E., & Hackett, G. (1981). The relationship of career related self-efficacy expectations to perceived career options in college women and men. *Journal of Counseling Psychology*, *42*, 22–26.

Flynn, D. T. (2016). STEM field persistence: The impact of engagement on postsecondary STEM persistence for underrepresented minority students. *Journal of Educational Issues*, *2*(1), 185–214. doi:10.5296/jei.v2i1.9245

Haller, E. J., & Virkler, S. J. (1993). Another look at rural-nonrural differences in students' educational aspirations. *Journal of Research in Rural Education*, *9*(3), 170–178.

Hayes, R. Q., Whalen, S. K., & Cannon, B. (2009). *2008–2009 CSRDE stem retention report*. Norman, OK: Center for Institutional Data Exchange and Analysis, University of Oklahoma.

Kassaee, A. M., & Rowell, G. H. (2016). Motivationally-Informed interventions for at-risk STEM students. *Journal of STEM Education: Innovations and Research*, *17*(3), 77–84.

Minnesota State Careerwise Education. (2017). *Understanding Science, Technology, Engineering, and Math (STEM) skills*. Retrieved from https://careerwise.minnstate.edu/careers/stemskills.html

National Center for Education Statistics. (2013). *The status of rural education*. Retrieved from: https://nces.ed.gov/programs/coe/indicator_tla.asp#info

Pajares, F., & Cheong, Y. F. (2003). Achievement goal orientations in writing: A developmental perspective. *International Journal of Educational Research*, *39*(4-5), 437–455. doi:10.1016/j.ijer.2004.06.008

Rosenzweig, E. Q., & Wigfield, A. (2016). STEM motivation interventions for adolescents: A promising start, but further to go. *Educational Psychologist*, *51*(2), 146–163. doi:10.1080/00461520.2016.1154792

Ryan, R. M., & Deci, E. L. (2000). Self-determination theory and the facilitation of intrinsic motivation, social development, and well-being. *The American Psychologist*, *55*(1), 68–78. doi:10.1037/0003-066X.55.1.68 PMID:11392867

Shin, J. L., Levy, S. R., & London, B. (2016). Effects of role model exposure on STEM and non-STEM student engagement. *Journal of Applied Social Psychology*, *410*(7). doi:10.1111/jasp.12371

Siekmann, G., & Korbel, P. (2016). *Defining 'STEM' skills: Review and synthesis of the literature — support document 2*. Adelaide: NCVER.

Strimel, G. (2014). Authentic education. *Technology & Engineering Teacher*, *73*(7), 8–18.

Strobel, J., Wang, J., Weber, N. R., & Dyehouse, M. (2013). The role of authenticity in design-based learning environments: The case of engineering education. *Computers & Education*, *64*, 143–152. doi:10.1016/j.compedu.2012.11.026

Tang, M., Pan, W., & Newmeyer, M. D. (2008). Factors influencing high school students' career aspirations. *Professional School Counseling*, *11*(5), 285–295. doi:10.5330/PSC.n.2010-11.285

Tinto, V. (1993). *Leaving college: Rethinking the causes and cures of student attrition* (2nd ed.). Chicago: University of Chicago Press.

Versypt, J. J., & Versypt, A. N. (2013). Mapping rural students' STEM involvement: Case studies of chemical engineering undergraduate enrollment in the states of Illinois and Kansas. *120th ASEE Annual Conference and Exposition*.

Wilson, Z. S., Holmes, L., deGravelles, K., Sylvain, M. R., Batiste, L., Johnson, M., ... Warner, I. M. (2012). Hierarchical mentoring: A transformative strategy for improving diversity and retention in undergraduate STEM disciplines. *Journal of Science Education and Technology*, *21*(1), 148–156. doi:10.100710956-011-9292-5

KEY TERMS AND DEFINITIONS

Authentic Learning: Hands-on activities where students learn by solving real world STEM problems.

Contextual Authenticity: Experiences in real world STEM environments.

Engaging: Capturing the attention of students.

Experiencing: Participating in STEM activities.

Experiential Learning: Learning acquired through hands-on, authentic experiences.

Impact Authenticity: Experiences in social contexts. Students realizing their potential to impact society through STEM professions.

Intrinsic Motivation: Performing an action or behavior for the sake of enjoyment.

More Knowledgeable Other (MKO): Someone or something having a better understanding or a higher ability level regarding a particular task, process, or concept. In the case of Project Engage, mentors, career counselors, and STEM faculty serve the role of MKOs for the undergraduate STEM majors.

Motivating: Stimulating interest in or enthusiasm for doing something.

Personal/Individual Authenticity: Integrating cultural values into students' learning experiences.

Project Engage: A comprehensive, capacity-building program which provides strategic mentoring, career exploration, as well as knowledge and skills to prepare undergraduate STEM majors, specifically minority students from rural areas for future STEM careers.

Scaffolding: Providing a framework for supporting student success.

STEM Self-Efficacy: A student's belief in his/her knowledge and skills related to success in science, technology, engineering, and/or mathematics.

Task Authenticity: Undertakings similar to STEM professionals. Provides students with the opportunity to develop professional STEM skills.

Zone of Proximal Development (ZPD): The level of cognitive development that the learner can obtain with the guidance of teachers or in collaboration with peers.

Related Readings

To continue IGI Global's long-standing tradition of advancing innovation through emerging research, please find below a compiled list of recommended IGI Global book chapters and journal articles in the areas of STEM education and science education. These related readings will provide additional information and guidance to further enrich your knowledge and assist you with your own research.

Akers, T., Nyarko, K., Gwynn, D., & Larkin, W. D. (2014). Visual Analytics for Students and Teachers (VAST) Model at a Minority Serving Institution: A Department of Homeland Security Project for Strengthening STEM Teaching and Learning. In C. Brown, K. Peters, & K. Nyarko (Eds.), *Cases on Research and Knowledge Discovery: Homeland Security Centers of Excellence* (pp. 152–169). Hershey, PA: IGI Global. doi:10.4018/978-1-4666-5946-9.ch007

Alkadi, I. (2016). Assessing Security with Regard to Cloud Applications in STEM Education. In L. Chao (Ed.), *Handbook of Research on Cloud-Based STEM Education for Improved Learning Outcomes* (pp. 260–276). Hershey, PA: IGI Global. doi:10.4018/978-1-4666-9924-3.ch017

Anohah, E. (2016). Pedagogy and Design of Online Learning Environment in Computer Science Education for High Schools. *International Journal of Online Pedagogy and Course Design*, *6*(3), 39–51. doi:10.4018/IJOPCD.2016070104

Anohah, E. (2017). Paradigm and Architecture of Computing Augmented Learning Management System for Computer Science Education. *International Journal of Online Pedagogy and Course Design*, *7*(2), 60–70. doi:10.4018/IJOPCD.2017040105

Anohah, E., & Suhonen, J. (2017). Trends of Mobile Learning in Computing Education from 2006 to 2014: A Systematic Review of Research Publications. *International Journal of Mobile and Blended Learning*, *9*(1), 16–33. doi:10.4018/IJMBL.2017010102

Arinze, B., Sylla, C., & Amobi, O. (2016). Cloud Computing for Teaching and Learning: Design Strategies. In L. Chao (Ed.), *Handbook of Research on Cloud-Based STEM Education for Improved Learning Outcomes* (pp. 159–171). Hershey, PA: IGI Global. doi:10.4018/978-1-4666-9924-3.ch011

Ault, M., Craig-Hare, J., Ellis, J. D., Bulgren, J., Kretschmer, I., & Frey, B. B. (2016). Using Reason Racer to Support Argumentation in Middle School Science Instruction. In M. Urban & D. Falvo (Eds.), *Improving K-12 STEM Education Outcomes through Technological Integration* (pp. 399–431). Hershey, PA: IGI Global. doi:10.4018/978-1-4666-9616-7.ch018

Back, S. M., Greenhalgh-Spencer, H., & Frias, K. M. (2016). The Application of Transdisciplinary Theory and Practice to STEM Education. In Y. Rosen, S. Ferrara, & M. Mosharraf (Eds.), *Handbook of Research on Technology Tools for Real-World Skill Development* (pp. 42–67). Hershey, PA: IGI Global. doi:10.4018/978-1-4666-9441-5.ch003

Barker, B. S., Nugent, G., Grandgenett, N., Keshwani, J., Nelson, C. A., & Leduc-Mills, B. (2016). Developing an Elementary Engineering Education Program through Problem-Based Wearable Technologies Activities. In J. Holland (Ed.), *Wearable Technology and Mobile Innovations for Next-Generation Education* (pp. 269–294). Hershey, PA: IGI Global. doi:10.4018/978-1-5225-0069-8.ch014

Bennett, A., Eglash, R., Lachney, M., & Babbitt, W. (2016). Design Agency: Diversifying Computer Science at the Intersections of Creativity and Culture. In M. Raisinghani (Ed.), *Revolutionizing Education through Web-Based Instruction* (pp. 35–56). Hershey, PA: IGI Global. doi:10.4018/978-1-4666-9932-8.ch003

Berdanier, C. G., Zephirin, T., Cox, M. F., & Black, S. M. (2015). Teaching MSE Students to Teach: A Design-Based Research Model for Introducing Professional Skills into the Technical Curriculum. In H. Lim (Ed.), *Handbook of Research on Recent Developments in Materials Science and Corrosion Engineering Education* (pp. 369–396). Hershey, PA: IGI Global. doi:10.4018/978-1-4666-8183-5.ch019

Bernhardt, S. (2014). Structural Context and Influencing Factors. In *Women in IT in the New Social Era: A Critical Evidence-Based Review of Gender Inequality and the Potential for Change* (pp. 127–157). Hershey, PA: IGI Global. doi:10.4018/978-1-4666-5860-8.ch006

Beziat, T. L., McCombs, K. M., Burks, B. A., & Byrom, J. (2016). Are Female STEM Majors Academic Risk Takers? In U. Thomas & J. Drake (Eds.), *Critical Research on Sexism and Racism in STEM Fields* (pp. 144–152). Hershey, PA: IGI Global. doi:10.4018/978-1-5225-0174-9.ch008

Brewley, D. N., Boindala, P. S., & Sinclair, J. L. (2015). Ideation to Execution: Flipping an Undergraduate Pre-Calculus Course to Create Significant Learning Experiences. In A. Scheg (Ed.), *Implementation and Critical Assessment of the Flipped Classroom Experience* (pp. 25–40). Hershey, PA: IGI Global. doi:10.4018/978-1-4666-7464-6.ch002

Brown, C. A. (2016). Using Digital Resources to Support STEM Education. In V. Wang (Ed.), *Handbook of Research on Learning Outcomes and Opportunities in the Digital Age* (pp. 127–151). Hershey, PA: IGI Global. doi:10.4018/978-1-4666-9577-1.ch006

Buck, G. A., Beeman-Cadwallader, N., & Trauth-Nare, A. (2016). Examining the Levels of Reasoning Used by Urban Elementary Black Girls Engaging in Technology-Enhanced Inquiry. In M. Urban & D. Falvo (Eds.), *Improving K-12 STEM Education Outcomes through Technological Integration* (pp. 86–107). Hershey, PA: IGI Global. doi:10.4018/978-1-4666-9616-7.ch005

Cai, T. (2015). From Learning Management System to Internet-Based Research in Mathematics Education. In A. Scheg (Ed.), *Critical Examinations of Distance Education Transformation across Disciplines* (pp. 62–85). Hershey, PA: IGI Global. doi:10.4018/978-1-4666-6555-2.ch004

Cartledge, D. (2014). Re–Valuing Constructed Knowledge: An Agenda for STE(A)M. In T. Fitzgerald (Ed.), *Advancing Knowledge in Higher Education: Universities in Turbulent Times* (pp. 170–186). Hershey, PA: IGI Global. doi:10.4018/978-1-4666-6202-5.ch011

Chang, S., von Imhoff, M., & Ustrup, R. I. (2016). Engineering and Information Technology: Challenges and Opportunities for Exchange Studies. In D. Velliaris & D. Coleman-George (Eds.), *Handbook of Research on Study Abroad Programs and Outbound Mobility* (pp. 300–324). Hershey, PA: IGI Global. doi:10.4018/978-1-5225-0169-5.ch012

Chapman, D. L., & Wang, S. (2015). Multimedia Instructional Tools and Student Learning in a Computer Applications Course. *International Journal of Information and Communication Technology Education, 11*(2), 57–67. doi:10.4018/ijicte.2015040105

Charles, D. R., & Wray, F. (2015). The English Science Cities: A New Phase in Science-based Urban Strategy. *International Journal of Knowledge-Based Organizations, 5*(1), 46–61. doi:10.4018/ijkbo.2015010104

Chen, Y. (2016). Security Issues Related to Cloud Applications in STEM Education. In L. Chao (Ed.), *Handbook of Research on Cloud-Based STEM Education for Improved Learning Outcomes* (pp. 277–289). Hershey, PA: IGI Global. doi:10.4018/978-1-4666-9924-3.ch018

Choi, H., & Feng, J. (2017). Using Video Games to Improve Spatial Skills. In R. Zheng & M. Gardner (Eds.), *Handbook of Research on Serious Games for Educational Applications* (pp. 93–114). Hershey, PA: IGI Global. doi:10.4018/978-1-5225-0513-6.ch005

Choi, J., An, S., & Lee, Y. (2015). A Case Study of Distance Education for Informatics Gifted Students. In L. Lennex & K. Nettleton (Eds.), *Cases on Instructional Technology in Gifted and Talented Education* (pp. 358–380). Hershey, PA: IGI Global. doi:10.4018/978-1-4666-6489-0.ch017

Coleman, N., Farina, D. M., & Rabinovich, L. (2017). Common Denominators to Learner-Centered Success: Undergraduate STEM, Graduate Teacher Education, and an Educational Technology Doctoral Program. In J. Keengwe & G. Onchwari (Eds.), *Handbook of Research on Learner-Centered Pedagogy in Teacher Education and Professional Development* (pp. 23–44). Hershey, PA: IGI Global. doi:10.4018/978-1-5225-0892-2.ch002

Coyle, V. C., Newman, D. L., & Connor, K. A. (2016). Innovative Instruction in STEM Education: The Role of Student Feedback in the Development of a Flipped Classroom. In J. Keengwe & G. Onchwari (Eds.), *Handbook of Research on Active Learning and the Flipped Classroom Model in the Digital Age* (pp. 309–332). Hershey, PA: IGI Global. doi:10.4018/978-1-4666-9680-8.ch016

Dass, P. M., & Spagnolo, J. T. (2016). STEM Learning in Middle Grades by Technology-Mediated Integration of Science and Mathematics: Results of Project SMILE. In M. Urban & D. Falvo (Eds.), *Improving K-12 STEM Education Outcomes through Technological Integration* (pp. 187–205). Hershey, PA: IGI Global. doi:10.4018/978-1-4666-9616-7.ch009

Dawson, M., Al Saeed, I., Wright, J., & Onyegbula, F. (2014). Open Source Software to Enhance the STEM Learning Environment. In V. Wang (Ed.), *Handbook of Research on Education and Technology in a Changing Society* (pp. 569–580). Hershey, PA: IGI Global. doi:10.4018/978-1-4666-6046-5.ch042

De la Garza, R. L., Zavala, G., & Gómez Galindo, A. A. (2017). Argumentation Schema to Analyze High School Students' Scientific Reasoning. In M. Ramírez-Montoya (Ed.), *Handbook of Research on Driving STEM Learning With Educational Technologies* (pp. 297–314). Hershey, PA: IGI Global. doi:10.4018/978-1-5225-2026-9.ch015

DeCoito, I., & Richardson, T. (2016). Using Technology to Enhance Science Literacy, Mathematics Literacy, or Technology Literacy: Focusing on Integrated STEM Concepts in a Digital Game. In M. Urban & D. Falvo (Eds.), *Improving K-12 STEM Education Outcomes through Technological Integration* (pp. 1–22). Hershey, PA: IGI Global. doi:10.4018/978-1-4666-9616-7.ch001

Delgado-Cepeda, F. J. (2016). Upgrading a Numerical Methods Course into New Mobile Technologies for Mathematical Education: An Approach Based on Flexibility and Skill Development. In D. Parsons (Ed.), *Mobile and Blended Learning Innovations for Improved Learning Outcomes* (pp. 201–222). Hershey, PA: IGI Global. doi:10.4018/978-1-5225-0359-0.ch011

Denner, J., Martinez, J., & Thiry, H. (2017). Strategies for Engaging Hispanic/Latino Youth in the US in Computer Science. In Y. Rankin & J. Thomas (Eds.), *Moving Students of Color from Consumers to Producers of Technology* (pp. 24–48). Hershey, PA: IGI Global. doi:10.4018/978-1-5225-2005-4.ch002

Deyoe, M. M., Newman, D. L., & Lamendola, J. M. (2015). Mobile Technology in Higher Education: Patterns of Replication and Transferability. In J. Keengwe & M. Maxfield (Eds.), *Advancing Higher Education with Mobile Learning Technologies: Cases, Trends, and Inquiry-Based Methods* (pp. 128–153). Hershey, PA: IGI Global. doi:10.4018/978-1-4666-6284-1.ch007

Dietz, G., Hessedence, J., Long, T., & Muga, H. E. (2017). Empowering Women in STEM: Embedding STEM in K-12 Education. In M. Gray & K. Thomas (Eds.), *Strategies for Increasing Diversity in Engineering Majors and Careers* (pp. 61–87). Hershey, PA: IGI Global. doi:10.4018/978-1-5225-2212-6.ch004

Dollahon, C. V. (2017). Using STEAM in Marine Science: Incorporating Graphic Design Into an Existing STEM Lesson. In J. Bazler & M. Van Sickle (Eds.), *Cases on STEAM Education in Practice* (pp. 292–317). Hershey, PA: IGI Global. doi:10.4018/978-1-5225-2334-5.ch015

Dongale, T. D., Patil, S. S., & Kamat, R. K. (2015). Learning by Simulations: A New and Effective Pedagogical Approach for Science, Engineering and Technology Students in a Traditional Setting. *International Journal of Quality Assurance in Engineering and Technology Education*, *4*(2), 13–25. doi:10.4018/IJQAETE.2015040102

Dragnic-Cindric, D., Barrow, E., & Anderson, J. L. (2017). Opportunity to Start Strong: Integration of Technology in Science Lessons in the Early Elementary Grades. In J. Keengwe & P. Bull (Eds.), *Handbook of Research on Transformative Digital Content and Learning Technologies* (pp. 154–170). Hershey, PA: IGI Global. doi:10.4018/978-1-5225-2000-9.ch009

Duran, A., & Lopez, D. (2015). Women from Diverse Backgrounds in the Science, Technology, Engineering, and Math (STEM) Professions: Retention and Career Development. In C. Hughes (Ed.), *Impact of Diversity on Organization and Career Development* (pp. 214–251). Hershey, PA: IGI Global. doi:10.4018/978-1-4666-7324-3.ch009

Dzan, W., Tsai, H., Lou, S., & Shih, R. (2015). Satisfaction Analysis of Experiential Learning-Based Popular Science Education. *International Journal of Distance Education Technologies*, *13*(2), 93–109. doi:10.4018/IJDET.2015040106

Eglash, R., Babbitt, W., Bennett, A., Bennett, K., Callahan, B., Davis, J., ... Tully, K. (2017). Culturally Situated Design Tools: Generative Justice as a Foundation for STEM Diversity. In Y. Rankin & J. Thomas (Eds.), *Moving Students of Color from Consumers to Producers of Technology* (pp. 132–151). Hershey, PA: IGI Global. doi:10.4018/978-1-5225-2005-4.ch007

Eguchi, A. (2015). Educational Robotics as a Learning Tool for Promoting Rich Environments for Active Learning (REALs). In J. Keengwe (Ed.), *Handbook of Research on Educational Technology Integration and Active Learning* (pp. 19–47). Hershey, PA: IGI Global. doi:10.4018/978-1-4666-8363-1.ch002

El-Farargy, N. (2015). Making Sense of Science: A Review in Scottish Further Education. In E. de Silva (Ed.), *Cases on Research-Based Teaching Methods in Science Education* (pp. 1–29). Hershey, PA: IGI Global. doi:10.4018/978-1-4666-6375-6.ch001

Erickson, J. J. (2015). To Play or to Learn?: A Review of Game-Based Math Learning for Motivation and Cognition. *International Journal of Cyber Behavior, Psychology and Learning*, *5*(1), 56–74. doi:10.4018/ijcbpl.2015010105

Estevez, C. (2016). Addressing Transport Layer Issues in Cloud Computing: A STEM Perspective. In L. Chao (Ed.), *Handbook of Research on Cloud-Based STEM Education for Improved Learning Outcomes* (pp. 79–93). Hershey, PA: IGI Global. doi:10.4018/978-1-4666-9924-3.ch006

Fakayode, S. O., Snipes, V., Kanipes, M. I., Mohammed, A. K., & Wilson, Z. S. (2016). Use of Innovative Pedagogies and Creative Partnership Strategies to Promote Undergraduate STEM Education at an HBCU. In C. Prince & R. Ford (Eds.), *Setting a New Agenda for Student Engagement and Retention in Historically Black Colleges and Universities* (pp. 119–151). Hershey, PA: IGI Global. doi:10.4018/978-1-5225-0308-8.ch008

Fakomogbon, M. A., Adebayo, R. F., Adegbija, M. V., Shittu, A. T., & Oyelekan, O. S. (2014). Subject Specialization and Science Teachers' Perception of Information and Communication Technology for Instruction in Kwara State. *International Journal of Information and Communication Technology Education*, *10*(1), 25–34. doi:10.4018/ijicte.2014010103

Farland-Smith, D., & Ledger, T. (2016). Understanding How Images and Attitudes toward Scientists and Science Contribute to Science Identities: Investigating How Images Drawn by Elementary, Middle, and High School Students Reflect Their Attitudes. In E. Railean, G. Walker, A. Elçi, & L. Jackson (Eds.), *Handbook of Research on Applied Learning Theory and Design in Modern Education* (pp. 498–518). Hershey, PA: IGI Global. doi:10.4018/978-1-4666-9634-1.ch024

Feldhaus, C. R., Little, J., & Sorge, B. (2015). Conflict Resolution and Ethical Decision-Making for Engineering Professionals in Global Organizations. In S. Sundar Sethy (Ed.), *Contemporary Ethical Issues in Engineering* (pp. 204–227). Hershey, PA: IGI Global. doi:10.4018/978-1-4666-8130-9.ch015

Fleenor, H. G., & Hodhod, R. (2016). Assessment of Learning and Technology: Computer Science Education. In V. Wang (Ed.), *Handbook of Research on Learning Outcomes and Opportunities in the Digital Age* (pp. 51–78). Hershey, PA: IGI Global. doi:10.4018/978-1-4666-9577-1.ch003

Fletcher, E. C. Jr, & Djajalaksana, Y. (2014). Predictors of Instructional Strategy Use of Faculty in Career and Technical Education Programs: Signature Pedagogies of the Field. *International Journal of Adult Vocational Education and Technology*, *5*(3), 45–65. doi:10.4018/ijavet.2014070104

Flournoy, B. (2016). Black Women and Science in Higher Education: Not Much Has Changed – Still the Struggle. In U. Thomas & J. Drake (Eds.), *Critical Research on Sexism and Racism in STEM Fields* (pp. 170–180). Hershey, PA: IGI Global. doi:10.4018/978-1-5225-0174-9.ch010

Gabriel, B. F., Valente, R. A., Dias-de-Oliveira, J., Neto, V. F., & Andrade-Campos, A. (2017). Methodologies for Engineering Learning and Teaching (MELT): An Overview of Engineering Education in Europe and a Novel Concept for Young Students. In L. Carvalho (Ed.), *Handbook of Research on Entrepreneurial Development and Innovation Within Smart Cities* (pp. 363–391). Hershey, PA: IGI Global. doi:10.4018/978-1-5225-1978-2.ch016

Galeshi, R. (2017). Engineering Education for All: Increasing Access to Engineering Education for Men and Women across the World through Distance Learning. *International Journal of Online Pedagogy and Course Design*, *7*(2), 35–47. doi:10.4018/IJOPCD.2017040103

Garner, N., Lischke, M. D., Siol, A., & Eilks, I. (2014). Learning about Sustainability in a Non-Formal Laboratory Context for Secondary Level Students: A Module on Climate Change, the Ozone Hole, and Summer Smog. In K. Thomas & H. Muga (Eds.), *Handbook of Research on Pedagogical Innovations for Sustainable Development* (pp. 229–244). Hershey, PA: IGI Global. doi:10.4018/978-1-4666-5856-1.ch012

Gatling, A. P. (2016). Technology's Role in Supporting Elementary Preservice Teachers as They Teach: An Urban STEM Afterschool Enrichment Program. In M. Urban & D. Falvo (Eds.), *Improving K-12 STEM Education Outcomes through Technological Integration* (pp. 362–379). Hershey, PA: IGI Global. doi:10.4018/978-1-4666-9616-7.ch016

Gomes, A. D., Mendes, A. J., & Marcelino, M. J. (2015). Computer Science Education Research: An Overview and Some Proposals. In R. Queirós (Ed.), *Innovative Teaching Strategies and New Learning Paradigms in Computer Programming* (pp. 1–29). Hershey, PA: IGI Global. doi:10.4018/978-1-4666-7304-5.ch001

Gorman, J., & Heinze-Fry, J. (2014). Conceptual Mapping Facilitates Coherence and Critical Thinking in the Science Education System. In L. Shedletsky & J. Beaudry (Eds.), *Cases on Teaching Critical Thinking through Visual Representation Strategies* (pp. 296–334). Hershey, PA: IGI Global. doi:10.4018/978-1-4666-5816-5.ch012

Günel, M., Özer-Keskin, M., & Keskin-Samancı, N. (2017). A Paradigm Shift for Teachers' Professional Development Structure in Turkey: Moving from Instruction to Learning. In *Educational Leadership and Administration: Concepts, Methodologies, Tools, and Applications* (pp. 282–302). Hershey, PA: IGI Global. doi:10.4018/978-1-5225-1624-8.ch016

Haswell, M. M. (2016). Barriers to Success: Disadvantages of Gender Normative Language to Women in STEM. In U. Thomas & J. Drake (Eds.), *Critical Research on Sexism and Racism in STEM Fields* (pp. 49–66). Hershey, PA: IGI Global. doi:10.4018/978-1-5225-0174-9.ch003

Havard, B., East, M. L., Prayaga, L., & Whiteside, A. (2016). Adaptable Learning Theory Framework for Technology-Enhanced Learning. In E. Railean, G. Walker, A. Elçi, & L. Jackson (Eds.), *Handbook of Research on Applied Learning Theory and Design in Modern Education* (pp. 632–654). Hershey, PA: IGI Global. doi:10.4018/978-1-4666-9634-1.ch031

Herr, N., Tippens, M., Rivas, M., Vandergon, V. O., d'Alessio, M., & Reveles, J. (2016). Cloud-Based Continuous Formative Assessment (CFA). In L. Chao (Ed.), *Handbook of Research on Cloud-Based STEM Education for Improved Learning Outcomes* (pp. 311–342). Hershey, PA: IGI Global. doi:10.4018/978-1-4666-9924-3.ch020

Hoban, G., Nielsen, W., & Hyland, C. (2016). Blended Media: Student-Generated Mash-ups to Promote Engagement with Science Content. *International Journal of Mobile and Blended Learning*, *8*(3), 35–48. doi:10.4018/IJMBL.2016070103

Hoge, B. (2014). GBL as PBL: Guidelines for Game-Based Learning in the Classroom and Informal Science Centers. In Z. Yang, H. Yang, D. Wu, & S. Liu (Eds.), *Transforming K-12 Classrooms with Digital Technology* (pp. 58–82). Hershey, PA: IGI Global. doi:10.4018/978-1-4666-4538-7.ch004

Jablon, P. C. (2017). Theater as the STEAM Engine for Engaging Those Previously Disengaged. In J. Bazler & M. Van Sickle (Eds.), *Cases on STEAM Education in Practice* (pp. 118–153). Hershey, PA: IGI Global. doi:10.4018/978-1-5225-2334-5.ch006

Jiang, E. P. (2014). A Comparative Study on Undergraduate Computer Science Education between China and the United States. In V. Wang (Ed.), *International Education and the Next-Generation Workforce: Competition in the Global Economy* (pp. 208–223). Hershey, PA: IGI Global. doi:10.4018/978-1-4666-4498-4.ch012

Johnson, J., Jones, T., Haddad, G., & Wilcox, C. (Gwathmey) Wilcox, J. K. (2016). Strategies to Enhance the Role of HBCUs in Increasing the Science, Technology, Engineering, Mathematics, and Medical (STEMM) Workforce. In C. Prince, & R. Ford (Eds.), Setting a New Agenda for Student Engagement and Retention in Historically Black Colleges and Universities (pp. 96-118). Hershey, PA: IGI Global. doi:10.4018/978-1-5225-0308-8.ch007

Johnson, M. L., & Spear, K. (2017). Developing a Learning Community of Engineers Through an Honors First-Year Seminar. In M. Gray, & K. Thomas (Eds.), *Strategies for Increasing Diversity in Engineering Majors and Careers* (pp. 88-105). Hershey, PA: IGI Global. 10.4018/978-1-5225-2212-6.ch005

Juanes, J. A., & Ruisoto, P. (2014). Technological Advances and Teaching Innovation Applied to Health Science Education. *Journal of Information Technology Research*, *7*(2), 1–6. doi:10.4018/jitr.2014040101

Kalaian, S. A., & Kasim, R. M. (2015). Small-Group vs. Competitive Learning in Computer Science Classrooms: A Meta-Analytic Review. In R. Queirós (Ed.), *Innovative Teaching Strategies and New Learning Paradigms in Computer Programming* (pp. 46–64). Hershey, PA: IGI Global. doi:10.4018/978-1-4666-7304-5.ch003

Karlsson, G. (2015). Students' Joint Reasoning about Gas Solubility in Water in Modified Versions of a Virtual Laboratory. *International Journal of Online Pedagogy and Course Design, 5*(4), 67–77. doi:10.4018/IJOPCD.2015100105

Kennedy, B. F., & McLaughlin, P. M. (2017). Growing Global STEM Learning in Higher Education. In L. Leavitt, S. Wisdom, & K. Leavitt (Eds.), *Cultural Awareness and Competency Development in Higher Education* (pp. 132–147). Hershey, PA: IGI Global. doi:10.4018/978-1-5225-2145-7.ch008

Kleanthous, I., & Meletiou-Mavrotheris, M. (2016). Early Statistical Reasoning: An Exploratory Study of Primary School Students' Use of a Dynamic Statistics Software Package for Analyzing and Interpreting Data. *International Journal of Information Communication Technologies and Human Development, 8*(1), 26–41. doi:10.4018/IJICTHD.2016010102

Komura, K. (2017). Professional Masters' STEM Graduate Education Programs to Develop a Business-Savvy Workforce. In B. Narasimharao, E. Wright, S. Prasad, & M. Joshi (Eds.), *Handbook of Research on Science Education and University Outreach as a Tool for Regional Development* (pp. 35–49). Hershey, PA: IGI Global. doi:10.4018/978-1-5225-1880-8.ch004

Kumar, V., & Sharma, D. (2017). Cloud Computing as a Catalyst in STEM Education. *International Journal of Information and Communication Technology Education, 13*(2), 38–51. doi:10.4018/IJICTE.2017040104

Kumtepe, A. T., & Genc-Kumtepe, E. (2014). STEM in Early Childhood Education: We Talk the Talk, But Do We Walk the Walk? In Z. Yang, H. Yang, D. Wu, & S. Liu (Eds.), *Transforming K-12 Classrooms with Digital Technology* (pp. 140–163). Hershey, PA: IGI Global. doi:10.4018/978-1-4666-4538-7.ch008

Kvasny, L., Trauth, E. M., & Joshi, K. D. (2016). The Role of HBCUs in Preparing African American Males for Careers in Information Technology. In C. Prince & R. Ford (Eds.), *Setting a New Agenda for Student Engagement and Retention in Historically Black Colleges and Universities* (pp. 234–250). Hershey, PA: IGI Global. doi:10.4018/978-1-5225-0308-8.ch013

Laffey, J. M., Sadler, T. D., Goggins, S. P., Griffin, J., & Babiuch, R. N. (2016). Mission HydroSci: Distance Learning through Game-Based 3D Virtual Learning Environments. In D. Russell & J. Laffey (Eds.), *Handbook of Research on Gaming Trends in P-12 Education* (pp. 421–441). Hershey, PA: IGI Global. doi:10.4018/978-1-4666-9629-7.ch020

Lang, C., Craig, A., & Egan, M. (2016). The Importance of Outreach Programs to Unblock the Pipeline and Broaden Diversity in ICT Education. *International Journal of Information and Communication Technology Education*, *12*(1), 38–49. doi:10.4018/IJICTE.2016010104

Leal, A. V., & Ferreira, D. J. (2016). Learning Programming Patterns Using Games. *International Journal of Information and Communication Technology Education*, *12*(2), 23–34. doi:10.4018/IJICTE.2016040103

Lee, Y., Dunbar, N., Kornelson, K., Wilson, S. N., Ralston, R., Savic, M., ... Elizondo, J. (2016). Digital Game based Learning for Undergraduate Calculus Education: Immersion, Calculation, and Conceptual Understanding. *International Journal of Gaming and Computer-Mediated Simulations*, *8*(1), 13–27. doi:10.4018/IJGCMS.2016010102

Lin, Y., Shen, M., & Liu, C. (2014). A Research of Employing Cognitive Load Theory in Science Education via Web-Pages. *International Journal of Online Pedagogy and Course Design*, *4*(2), 19–34. doi:10.4018/ijopcd.2014040102

Liu, X., Xu, C., & Ouyang, B. (2015). Plagiarism Detection Algorithm for Source Code in Computer Science Education. *International Journal of Distance Education Technologies*, *13*(4), 29–39. doi:10.4018/IJDET.2015100102

Lopez, C. (2015). Community Colleges and the Global Workforce: The Role and Impact of Community Colleges on the Pursuit of STEM Education and the Workforce. In S. Jones & D. Smith (Eds.), *Examining the Impact of Community Colleges on the Global Workforce* (pp. 209–228). Hershey, PA: IGI Global. doi:10.4018/978-1-4666-8481-2.ch011

Lou, S., Tsai, H., Tseng, K., & Shih, R. (2014). Effects of Implementing STEM-I Project-Based Learning Activities for Female High School Students. *International Journal of Distance Education Technologies*, *12*(1), 52–73. doi:10.4018/ijdet.2014010104

Ma, J., Chiu, D. K., & Tang, J. K. (2016). Exploring the Use of Social Media to Advance K12 Science Education. *International Journal of Systems and Service-Oriented Engineering*, *6*(4), 47–59. doi:10.4018/IJSSOE.2016100104

Mai, M. Y., Halim, L., Yaseen, R. M., & Meerah, T. S. (2014). Science, Technology, and Society (STS) Issues: A Survey of Secondary Students in Sana'a, Yemen. In M. Khosrow-Pour (Ed.), *Educational Technology Use and Design for Improved Learning Opportunities* (pp. 203–212). Hershey, PA: IGI Global. doi:10.4018/978-1-4666-6102-8.ch011

Marrero, M. E., Gunning, A. M., & Woodruff, K. (2016). Using Authentic Earth Data in the K-12 Classroom. In M. Urban & D. Falvo (Eds.), *Improving K-12 STEM Education Outcomes through Technological Integration* (pp. 281–309). Hershey, PA: IGI Global. doi:10.4018/978-1-4666-9616-7.ch013

Marshall, J. A., Erickson, T., & Sivam, K. (2015). Preservice Teachers Exploring the Nature of Science in Simulated Worlds. *International Journal of Gaming and Computer-Mediated Simulations*, *7*(2), 24–45. doi:10.4018/IJGCMS.2015040102

Martin, S. F., Green, A., & Dean, M. (2016). African American Women in STEM Education: The Cycle of Microaggressions from P-12 Classrooms to Higher Education and Back. In U. Thomas & J. Drake (Eds.), *Critical Research on Sexism and Racism in STEM Fields* (pp. 135–143). Hershey, PA: IGI Global. doi:10.4018/978-1-5225-0174-9.ch007

Martinez-Garza, M. M. (2015). Examining Epistemic Practices of the Community of Players of Dwarf Fortress: "For!! SCIENCE!! *International Journal of Gaming and Computer-Mediated Simulations*, *7*(2), 46–67. doi:10.4018/IJGCMS.2015040103

Mazuritskiy, M., Safontsev, S., Konoplev, B., & Boldyreva, A. (2014). Remote Access to Scientific Laboratory Equipment and Competency-Based Approach to Science and Technology Education. *International Journal of Information and Communication Technology Education*, *10*(3), 19–33. doi:10.4018/ijicte.2014070102

Moallem, M., Morge, S. P., Narayan, S., & Tagliarini, G. A. (2016). The Power of Computational Modeling and Simulation for Learning STEM Content in Middle and High Schools. In M. Urban & D. Falvo (Eds.), *Improving K-12 STEM Education Outcomes through Technological Integration* (pp. 135–171). Hershey, PA: IGI Global. doi:10.4018/978-1-4666-9616-7.ch007

Mpofu, C. (2016). STEM Fields and Ethnic Women in New Zealand: Issues of Sexism and Racism. In U. Thomas & J. Drake (Eds.), *Critical Research on Sexism and Racism in STEM Fields* (pp. 181–207). Hershey, PA: IGI Global. doi:10.4018/978-1-5225-0174-9.ch011

Mtebe, J. S., & Kissaka, M. M. (2016). Enhancing the Quality of Computer Science Education with MOOCs in Sub-Saharan Africa. In J. Keengwe & G. Onchwari (Eds.), *Handbook of Research on Active Learning and the Flipped Classroom Model in the Digital Age* (pp. 366–377). Hershey, PA: IGI Global. doi:10.4018/978-1-4666-9680-8.ch019

Mulvenon, S. W., & Wang, V. C. (2015). Graduate Level Research Methods and Statistics Courses: The Perspective of an Instructor. *International Journal of Adult Vocational Education and Technology*, *6*(4), 1–11. doi:10.4018/IJAVET.2015100101

Mungai, J. N. (2015). Quality Preparation of Mathematics and Science Teachers to Integrate ICT: Lessons from Learner-Centered Teacher Professional Development Approach. In F. Nafukho & B. Irby (Eds.), *Handbook of Research on Innovative Technology Integration in Higher Education* (pp. 29–53). Hershey, PA: IGI Global. doi:10.4018/978-1-4666-8170-5.ch002

Neri de Souza, F. (2016). Science Education with and through ICT: Curriculum Design and Questioning to Promote Active Learning. In D. Fonseca & E. Redondo (Eds.), *Handbook of Research on Applied E-Learning in Engineering and Architecture Education* (pp. 1–14). Hershey, PA: IGI Global. doi:10.4018/978-1-4666-8803-2.ch001

Newman, D. L., Deyoe, M. M., Connor, K. A., & Lamendola, J. M. (2014). Flipping STEM Learning: Impact on Students' Process of Learning and Faculty Instructional Activities. In J. Keengwe, G. Onchwari, & J. Oigara (Eds.), *Promoting Active Learning through the Flipped Classroom Model* (pp. 113–131). Hershey, PA: IGI Global. doi:10.4018/978-1-4666-4987-3.ch006

Newman, D. L., Lamendola, J. M., Deyoe, M. M., & Connor, K. A. (2015). Active Learning, Mentoring, and Mobile Technology: Meeting Needs across Levels in One Place. In J. Keengwe (Ed.), *Promoting Active Learning through the Integration of Mobile and Ubiquitous Technologies* (pp. 116–134). Hershey, PA: IGI Global. doi:10.4018/978-1-4666-6343-5.ch007

Ojeda-Castro, A. M., Murray-Finley, P., & Sánchez-Villafañe, J. (2017). Learning Management System Use to Increase Mathematics Knowledge and Skills in Puerto Rico. *International Journal of Technology and Human Interaction*, *13*(2), 89–100. doi:10.4018/IJTHI.2017040106

Ojokoh, B. A., Adeola, O. S., Isinkaye, F. O., & Abraham, C. (2014). Career Choices in Information and Communication Technology among South Western Nigerian Women. *Journal of Global Information Management*, *22*(2), 48–77. doi:10.4018/jgim.2014040104

Osisioma, I. U. (2017). Africanizing Science Education: Engaging Students in Context-Based Science Instruction. In R. Gordon, T. Akutsu, J. McDermott, & J. Lalas (Eds.), *Challenges Associated with Cross-Cultural and At-Risk Student Engagement* (pp. 120–147). Hershey, PA: IGI Global. doi:10.4018/978-1-5225-1894-5.ch007

Oyelekan, O. S., Akinpelu, G. A., & Daramola, F. O. (2015). Science Students' Use of the Internet for Learning in Higher Institutions in Osun State, Nigeria. *International Journal of Information and Communication Technology Education*, *11*(4), 67–82. doi:10.4018/IJICTE.2015100105

Peters, K. A., Wright Brown, C., & Nyarko, K. (2014). Conclusion: Perspectives on Effective Collaborative STEM Research Experiences Linked to DHS Centers of Excellence (COE). In C. Brown, K. Peters, & K. Nyarko (Eds.), *Cases on Research and Knowledge Discovery: Homeland Security Centers of Excellence* (pp. 268–292). Hershey, PA: IGI Global. doi:10.4018/978-1-4666-5946-9.ch012

Popov, O. (2015). Outdoor Science in Teacher Education. In T. Hansson (Ed.), *Contemporary Approaches to Activity Theory: Interdisciplinary Perspectives on Human Behavior* (pp. 128–142). Hershey, PA: IGI Global. doi:10.4018/978-1-4666-6603-0.ch008

Pugh, V. M. (2014). Lessons from the STEM Sector. In J. Prescott & J. McGurren (Eds.), *Gender Considerations and Influence in the Digital Media and Gaming Industry* (pp. 175–185). Hershey, PA: IGI Global. doi:10.4018/978-1-4666-6142-4.ch009

Qian, Y. (2015). 3D Multi-User Virtual Environments in Science Education: Potential and Challenges. In T. Heafner, R. Hartshorne, & T. Petty (Eds.), *Exploring the Effectiveness of Online Education in K-12 Environments* (pp. 302–324). Hershey, PA: IGI Global. doi:10.4018/978-1-4666-6383-1.ch015

Rappaport, J. M., Richter, S. B., & Kennedy, D. T. (2016). A Strategic Perspective on Using Symbolic Transformation in STEM Education: Robotics and Automation. *International Journal of Strategic Decision Sciences*, *7*(1), 39–75. doi:10.4018/IJSDS.2016010103

Redmond, A., Smith, B., & Smith, D. (2014). The Curriculum Development of a BIM Resilience Program for the National Institute of Building Science Facility Module. *International Journal of 3-D Information Modeling, 3*(1), 49-60. doi:10.4018/ij3dim.2014010105

Rickards, T. (2016). Improving the Effectiveness of Research Supervision in STEM Education: Cloud-Based Multimedia Solutions. In L. Chao (Ed.), *Handbook of Research on Cloud-Based STEM Education for Improved Learning Outcomes* (pp. 343–356). Hershey, PA: IGI Global. doi:10.4018/978-1-4666-9924-3.ch021

Rickards, T., & Steele, A. (2016). Designing a Cloud-Based Assessment Model: A New Zealand Polytechnic Case Study. In L. Chao (Ed.), *Handbook of Research on Cloud-Based STEM Education for Improved Learning Outcomes* (pp. 357–376). Hershey, PA: IGI Global. doi:10.4018/978-1-4666-9924-3.ch022

Rizzardini, R. H., & Gütl, C. (2016). A Cloud-Based Learning Platform: STEM Learning Experiences with New Tools. In L. Chao (Ed.), *Handbook of Research on Cloud-Based STEM Education for Improved Learning Outcomes* (pp. 106–122). Hershey, PA: IGI Global. doi:10.4018/978-1-4666-9924-3.ch008

Ro, H. K., & McIntosh, K. (2016). Constructing Conducive Environment for Women of Color in Engineering Undergraduate Education. In U. Thomas & J. Drake (Eds.), *Critical Research on Sexism and Racism in STEM Fields* (pp. 23–48). Hershey, PA: IGI Global. doi:10.4018/978-1-5225-0174-9.ch002

Roberts, K. L. (2015). Interdisciplinary Course Development in Nanostructured Materials Science and Engineering. In H. Lim (Ed.), *Handbook of Research on Recent Developments in Materials Science and Corrosion Engineering Education* (pp. 212–229). Hershey, PA: IGI Global. doi:10.4018/978-1-4666-8183-5.ch011

Rogers, R. R., Winship, J., & Sun, Y. (2016). Systematic Support for STEM Pre-Service Teachers: An Authentic and Sustainable Four-Pillar Professional Development Model. In K. Dikilitaş (Ed.), *Innovative Professional Development Methods and Strategies for STEM Education* (pp. 73–90). Hershey, PA: IGI Global. doi:10.4018/978-1-4666-9471-2.ch005

Russo, M. R., & Brittain, K. (2014). Women and STEM. In V. Wang (Ed.), *Handbook of Research on Education and Technology in a Changing Society* (pp. 703–713). Hershey, PA: IGI Global. doi:10.4018/978-1-4666-6046-5.ch052

Sahlin, J. P., & Lobera, K. (2016). Cloud Computing as a Catalyst for Change in STEM Education. In L. Chao (Ed.), *Handbook of Research on Cloud-Based STEM Education for Improved Learning Outcomes* (pp. 12–30). Hershey, PA: IGI Global. doi:10.4018/978-1-4666-9924-3.ch002

Sakulich, A., El-Korchi, T., & Sisson, R. D. Jr. (2015). The Interdisciplinary, Project-Based Infrastructure Degradation Curriculum at Worcester Polytechnic Institute. In H. Lim (Ed.), *Handbook of Research on Recent Developments in Materials Science and Corrosion Engineering Education* (pp. 67–81). Hershey, PA: IGI Global. doi:10.4018/978-1-4666-8183-5.ch004

Sakulich, A., & Peterson, A. (2017). A Globally Focused, Experiential Educational System for STEM Fields: Measures for Intentionally Promoting Diversity. In M. Gray & K. Thomas (Eds.), *Strategies for Increasing Diversity in Engineering Majors and Careers* (pp. 176–200). Hershey, PA: IGI Global. doi:10.4018/978-1-5225-2212-6.ch009

Sciffer, S., & Shah, M. (2015). Widening the Participation of Disadvantaged Students in Engineering. *International Journal of Quality Assurance in Engineering and Technology Education*, *4*(1), 1–13. doi:10.4018/IJQAETE.2015010101

Seals, C., Horton, A., Berzina-Pitcher, I., & Mishra, P. (2017). A New Understanding of our Confusion: Insights from a Year-Long STEM Fellowship Program. In C. Martin & D. Polly (Eds.), *Handbook of Research on Teacher Education and Professional Development* (pp. 582–604). Hershey, PA: IGI Global. doi:10.4018/978-1-5225-1067-3.ch032

Sharpe, R. V. (2016). HBCUs: Efficiencies of Creating a Scientific Workforce Outta Fifteen Cents. In C. Prince & R. Ford (Eds.), *Administrative Challenges and Organizational Leadership in Historically Black Colleges and Universities* (pp. 199–212). Hershey, PA: IGI Global. doi:10.4018/978-1-5225-0311-8.ch011

Smith, L. K., Rooney-Varga, J. N., Gold, A. U., Oonk, D. J., & Morrison, D. (2016). Media Literacy as a Pathway to Bridge the Digital and STEM Divides: Interest Driven Media Projects for Teachers in the Trenches. In M. Urban & D. Falvo (Eds.), *Improving K-12 STEM Education Outcomes through Technological Integration* (pp. 23–43). Hershey, PA: IGI Global. doi:10.4018/978-1-4666-9616-7.ch002

Smith, P., Fulton, J., Irons, A., & Sanders, G. (2016). The Role of the Professional Doctorate in Developing Professional Practice in STEM Subjects. In K. Dikilitaş (Ed.), *Innovative Professional Development Methods and Strategies for STEM Education* (pp. 1–16). Hershey, PA: IGI Global. doi:10.4018/978-1-4666-9471-2.ch001

Smith-Walters, C., & Barker, H. L. (2015). Pre-Service Teachers' Self-Efficacy and Attitudes toward Learning and Teaching Science in a Content Course. In E. de Silva (Ed.), *Cases on Research-Based Teaching Methods in Science Education* (pp. 277–302). Hershey, PA: IGI Global. doi:10.4018/978-1-4666-6375-6.ch012

Srisawasdi, N. (2016). Motivating Inquiry-Based Learning Through a Combination of Physical and Virtual Computer-Based Laboratory Experiments in High School Science. In M. Urban & D. Falvo (Eds.), *Improving K-12 STEM Education Outcomes through Technological Integration* (pp. 108–134). Hershey, PA: IGI Global. doi:10.4018/978-1-4666-9616-7.ch006

Sung, W., Ahn, J., Kai, S. M., Choi, A., & Black, J. B. (2016). Incorporating Touch-Based Tablets into Classroom Activities: Fostering Children's Computational Thinking through iPad Integrated Instruction. In D. Mentor (Ed.), *Handbook of Research on Mobile Learning in Contemporary Classrooms* (pp. 378–406). Hershey, PA: IGI Global. doi:10.4018/978-1-5225-0251-7.ch019

Swaid, S. I. (2014). A Novel Strategy to Improve STEM Education: The E-Science Approach. In V. Wang (Ed.), *Handbook of Research on Education and Technology in a Changing Society* (pp. 860–871). Hershey, PA: IGI Global. doi:10.4018/978-1-4666-6046-5.ch064

Sweeny, R. W. (2017). The Politics of Video Games in STEM Education. In R. Shin (Ed.), *Convergence of Contemporary Art, Visual Culture, and Global Civic Engagement* (pp. 331–341). Hershey, PA: IGI Global. doi:10.4018/978-1-5225-1665-1.ch019

Taber, K. S. (2016). Constructivism in Education: Interpretations and Criticisms from Science Education. In E. Railean, G. Walker, A. Elçi, & L. Jackson (Eds.), *Handbook of Research on Applied Learning Theory and Design in Modern Education* (pp. 116–144). Hershey, PA: IGI Global. doi:10.4018/978-1-4666-9634-1.ch006

Taber, K. S., Ruthven, K., Howe, C., Mercer, N., Riga, F., Hofmann, R., & Luthman, S. (2015). Developing a Research-Informed Teaching Module for Learning about Electrical Circuits at Lower Secondary School Level: Supporting Personal Learning about Science and the Nature of Science. In E. de Silva (Ed.), *Cases on Research-Based Teaching Methods in Science Education* (pp. 122–156). Hershey, PA: IGI Global. doi:10.4018/978-1-4666-6375-6.ch006

Tang, K. C., Chiu, D. K., Chan, A. W., & Tang, J. K. (2016). Enterprise Content Management System for IT Training. *International Journal of Systems and Service-Oriented Engineering*, *6*(3), 32–54. doi:10.4018/IJSSOE.2016070103

Tang, M. S., & Karunanithi, A. T. (2016). A Media Interpretation of the Negative Impact of Cloud Technologies. In L. Chao (Ed.), *Handbook of Research on Cloud-Based STEM Education for Improved Learning Outcomes* (pp. 403–420). Hershey, PA: IGI Global. doi:10.4018/978-1-4666-9924-3.ch024

Tavernise, A., Bertacchini, F., Pantano, P. S., & Bilotta, E. (2015). Implementing a New Class-Lab: Guidelines for Integrating Innovative Devices in Pre-Service Teachers' Practice. *International Journal of Digital Literacy and Digital Competence*, *6*(3), 33–49. doi:10.4018/IJDLDC.2015070103

Terrazas-Arellanes, F. E., Knox, C., Strycker, L. A., & Walden, E. D. (2017). Online Learning Tools for Middle School Science: Lessons Learned from a Design-Based Research Project. *International Journal of Information and Communication Technology Education*, *13*(1), 27–40. doi:10.4018/IJICTE.2017010103

Terrazas-Arellanes, F. E., Knox, C., & Walden, E. (2015). Pilot Study on the Feasibility and Indicator Effects of Collaborative Online Projects on Science Learning for English Learners. *International Journal of Information and Communication Technology Education*, *11*(4), 31–50. doi:10.4018/IJICTE.2015100103

Thien, N. D., Terracina, A., Iocchi, L., & Mecella, M. (2016). Robotic Teaching Assistance for the "Tower of Hanoi" Problem. *International Journal of Distance Education Technologies*, *14*(1), 64–76. doi:10.4018/IJDET.2016010104

Thomas, U., & Drake, J. (2016). Grace under Fire: Examining the Underrepresentation and Persistence of Women in STEM Professions from an Ecological Systems Theory Perspective. In U. Thomas & J. Drake (Eds.), *Critical Research on Sexism and Racism in STEM Fields* (pp. 1–22). Hershey, PA: IGI Global. doi:10.4018/978-1-5225-0174-9.ch001

Ursyn, A. (2016). Teaching and Learning Science as a Visual Experience. In A. Ursyn (Ed.), *Knowledge Visualization and Visual Literacy in Science Education* (pp. 1–27). Hershey, PA: IGI Global. doi:10.4018/978-1-5225-0480-1.ch001

Van Sickle, M., & Koester, M. (2017). Musing on Unanswered Questions. In J. Bazler & M. Van Sickle (Eds.), *Cases on STEAM Education in Practice* (pp. 1–20). Hershey, PA: IGI Global. doi:10.4018/978-1-5225-2334-5.ch001

Vandergon, V. O., Reveles, J., Herr, N., Nguyen-Graf, D., Rivas, M., d'Alessio, M., & Foley, B. (2016). Engaging Students in Conducting Data Analysis: The Whole-Class Data Advantage. In L. Chao (Ed.), *Handbook of Research on Cloud-Based STEM Education for Improved Learning Outcomes* (pp. 172–197). Hershey, PA: IGI Global. doi:10.4018/978-1-4666-9924-3.ch012

Vebrianto, R., Rery, R. U., & Osman, K. (2016). BIOMIND Portal for Developing 21st Century Skills and Overcoming Students' Misconception in Biology Subject. *International Journal of Distance Education Technologies*, *14*(4), 55–67. doi:10.4018/IJDET.2016100105

Velázquez-Iturbide, J. Á., Debdi, O., & Paredes-Velasco, M. (2015). A Review of Teaching and Learning through Practice of Optimization Algorithms. In R. Queirós (Ed.), *Innovative Teaching Strategies and New Learning Paradigms in Computer Programming* (pp. 65–87). Hershey, PA: IGI Global. doi:10.4018/978-1-4666-7304-5.ch004

Webb, A. M., Knight, S. L., Ben Wu, X., & Schielack, J. F. (2014). Teaching Science with Web-Based Inquiry Projects: An Exploratory Investigation. *International Journal of Virtual and Personal Learning Environments*, *5*(2), 57–68. doi:10.4018/ijvple.2014040105

Weber, A. S. (2016). Strategic Planning for Cloud Computing Adoption in STEM Education: Finding Best Practice Solutions. In L. Chao (Ed.), *Handbook of Research on Cloud-Based STEM Education for Improved Learning Outcomes* (pp. 1–11). Hershey, PA: IGI Global. doi:10.4018/978-1-4666-9924-3.ch001

Williams, P. H., & Lee, C. K. (2017). Collaborations between Elementary Schools and Higher Education: Lessons Learned from Community-Engaged Science Education Efforts. In C. Crosby & F. Brockmeier (Eds.), *Community Engagement Program Implementation and Teacher Preparation for 21st Century Education* (pp. 82–104). Hershey, PA: IGI Global. doi:10.4018/978-1-5225-0871-7.ch005

Xia, B. S. (2017). A Pedagogical Review of Programming Education Research: What Have We Learned. *International Journal of Online Pedagogy and Course Design*, *7*(1), 33–42. doi:10.4018/IJOPCD.2017010103

Yantz, J., Smith, B. D., Rowell, G. H., Cheatham, T., Nelson, D., Stephens, D. C., & Tenpenny, E. B. (2015). Using Multidisciplinary Research Experiences to Enhance STEM Learning through Undergraduate, Team-Based, Summer Research Projects for At-Risk Students. In E. de Silva (Ed.), *Cases on Research-Based Teaching Methods in Science Education* (pp. 198–220). Hershey, PA: IGI Global. doi:10.4018/978-1-4666-6375-6.ch008

Yu, L. (2016). Experience Teaching Introductory Programming Courses. *International Journal of Quality Assurance in Engineering and Technology Education*, *5*(1), 40–53. doi:10.4018/IJQAETE.2016010104

Zaharias, P., Chatzeparaskevaidou, I., & Karaoli, F. (2017). Learning Geography Through Serious Games: The Effects of 2-Dimensional and 3-Dimensional Games on Learning Effectiveness, Motivation to Learn and User Experience. *International Journal of Gaming and Computer-Mediated Simulations*, *9*(1), 28–44. doi:10.4018/IJGCMS.2017010102

About the Authors

Reenay Rogers, an associate professor in Educational Research, holds a Ph.D. in Instructional Leadership with an emphasis in Instructional Technology and a M.A. in Secondary Science Education. As a former high school science teacher with over ten years of experience, Dr. Rogers recognizes the need for hands-on experiential learning and real world applications of STEM in the classroom. Dr. Rogers' experience as principal investigator on two Minority Science and Engineering Improvement Program grants has also made her uniquely aware of the importance of preparing women and minorities for future STEM careers. Rural education and STEM education are both important aspects of Dr. Rogers' research agenda. She has co-authored two book chapters; one on pre-service teacher preparation in STEM and one on STEM grant writing as well as conducted numerous presentations on technology for instructional purposes, STEM and rural education preparation, as well as rural instructional leader preparation.

Yan Sun is an Assistant Professor in Instructional Technology at Mississippi State University. She received her Ph.D. degree in Learning, Design, & Technology from Purdue University and completed her post-doctoral research work at Texas A&M University. Dr. Sun's research and teaching revolve around the area where STEM education intersects with technology. She is particularly interested in examining how emerging technology can be integrated to improve STEM education and to enhance students' interest and motivation in learning STEM and pursuing STEM careers. Dr. Sun has expertise in quantitative and mixed-methods research, and has been applying quantitative and mixed-methods methodologies in her research on innovative technology-integrated STEM education pedagogy, projects, and interventions.

Index

A

Alabama Nature Center 69-70, 92-95, 112, 119, 123, 168, 171
authentic learning 28, 85, 128, 137, 167, 170-171, 180

B

Black Belt 1-2, 10-12, 20, 23, 25-27, 29, 31, 33-34, 84, 168, 176

C

career counselor 30, 39-40, 44-45, 47-49, 60, 62-63, 86, 127, 131, 167, 174-175
career decidedness 120-123, 127-128, 157, 162
career exploration 13, 20-22, 25, 27, 29-32, 34, 38, 67, 70, 77, 115, 119-127, 137, 157-158, 162, 164, 167-168, 181
career self-efficacy 121-122, 127, 157, 162, 166
career thoughts 120-123, 126-128, 157-158, 162
cognitive development theory 40-41, 63
Community Mentoring 44, 63
competitive STEM events 30, 44, 77, 119, 164, 167
Contextual Authenticity 172, 180

E

educational preparation 25, 38, 42, 66
engaging 3, 23, 25, 28-31, 73, 83, 120, 131, 164, 170-171, 173-175, 180
experiencing 23, 25, 28-29, 67, 76, 164, 170-171, 175, 180
Experiential Learning 180

G

graduate level career counselor 45, 60, 63
guided career exploration 13, 21-22, 25, 29, 31, 34, 38, 120-123, 125, 127, 137, 162

H

Hierarchical Mentoring 44, 63
high school outreach 162, 170
Hudson Alpha Institute for Biotechnology 119
HudsonAlpha 74, 123, 171

I

Impact Authenticity 172-173, 180
Intrinsic Motivation 28, 166, 176, 180
Invention Convention 30, 68, 77-79, 82, 84, 86, 88, 103-105, 107, 114, 119, 128-133, 141-149, 158, 162, 164, 167

M

Marshall Space Flight 69, 71-73, 79-80, 83, 103, 119, 123, 171-172
Marshall Space Flight Center 69, 71-72, 79-80, 83, 103, 119, 123, 171-172
Mercedes-Benz 69, 75-76, 96, 119
Mercedes-Benz U.S. International

Incorporated 96, 119
Minority Science and Engineering Improvement Program 2, 11, 20, 38
minority students 1-2, 10-11, 14, 20, 22, 25-26, 31, 38, 49, 58, 64, 126, 128, 157, 165, 181
More Knowledgeable Other (MKO) 41, 59, 63, 174, 181
motivating 23, 25, 28-29, 129, 164, 166-168, 170, 176, 181
motivation 13, 21, 25, 28, 30, 38, 40, 42-43, 46, 59, 65-66, 72, 76-77, 79, 82, 85, 134, 163, 165-166, 171, 174, 176, 180

P

Peer Mentor 40, 46, 48, 63
Personal 69, 109, 174, 181
personal authenticity 174
Professional Exploration Trips 68, 70, 76, 119, 164
Project Engage 1-2, 10-14, 20-25, 27-35, 38-39, 42-53, 55, 57-59, 63-70, 76-81, 83-88, 94, 97-98, 108-109, 115, 120-132, 135-137, 141, 156-158, 162-165, 167-178, 181

R

Redstone Arsenal 69, 71-73, 119
rural schools 4-5, 7-11, 20, 84, 130, 135, 158, 165, 173
rural students 1-2, 5-9, 11-14, 20-23, 25, 39, 47, 49, 53, 58, 65, 68, 80, 83, 85, 120-121, 136, 158, 163, 165-166, 170, 175, 177

S

scaffolding 23, 25, 28-29, 39-44, 48-49, 59, 163-164, 170, 174-175, 177, 181
Sisters of the Academy 126, 162, 164
social constructivism 40, 42, 44, 63
STEM Challenge 30, 68, 77, 80-82, 85, 99, 101-103, 113, 119, 125, 129, 134-136, 150, 154-158, 162, 164, 167, 173
STEM engagement 12, 21-22, 25, 29-30, 38, 44, 73
STEM experiences 14, 21, 23, 25-26, 38, 163, 170
STEM pipeline 1-2, 4, 7, 9, 12, 20-21, 31, 57, 77, 114, 164-165, 170, 177-178
STEM role models 21, 27, 29-30, 38, 65, 168
STEM self-efficacy 30, 45, 59, 80, 119, 126, 128, 158, 166-167, 175, 181
STEM speakers 44, 68, 83, 85, 115, 119, 164, 168
strategic mentoring 20-22, 25, 29-31, 34, 38-39, 42-43, 46, 50, 59, 120, 164, 167, 181

T

Task Authenticity 172-173, 181
team building 31-32, 68, 73, 85-86, 88, 108-109, 114, 119, 164, 170

U

U.S. Space and Rocket Center 69, 88, 119
underfunded 20, 135
underrepresented groups 11, 14, 20, 22, 115, 165

V

vocational maturity 120-123, 127, 157-158, 162

Z

Zone of Proximal Development (ZPD) 39, 41, 59, 63, 181

www.ingramcontent.com/pod-product-compliance
Ingram Content Group UK Ltd.
Pitfield, Milton Keynes, MK11 3LW, UK
UKHW050945290926
15470UKWH00005B/1645

9 781522 563419